Financial Aspects in Agriculture

Indian Council of
Agricultural Research (ICAR)

Dr. Balasaheb Sawant Konkan
Krishi Vidyapeeth (DBSKKV)

National Agricultural Higher Education
Project-Institutional Development Plan
(NAHEP-IDP)

NIPA® Genx Electronic Resources &
Solutions P. Ltd., New Delhi-110 034

About the Authors

CMA Dr. Kinnarry Thakkar is Professor, Department of Commerce, University of Mumbai. Her qualifications are CMA, CS., Ph.D, MBA (Finance) and M.Com (Gold medalist). She has 25 years of teaching experience. She was Head- Department of Commerce, University of Mumbai during the period February 2019 to February 2022. She was Principal of SIES College of Commerce and Economics, Sion, Mumbai. Her area of specialisation is Finance and Accountancy.

She has 4 patents granted to her credit.

1. She is receipient of George Oliver International Distinguished Professor and Researcher Award 2023 on UN Public Service Day from IMRF 2023
2. Pride of Maharashtra Award for outstanding contribution in field of Academics from Navyug Arts & Commerce College, Jabalpur 2023
3. Eminent Academic Leader Award from I2OR- A registered MSME with Ministry of MSME, Govt of India
4. Dr. APJ Abdul Kalam Award for Teaching Excellence 2018.
5. Shiksha Ratna Puraskar for meritorious performance in Teaching

She is member, Board of Studies in University of Mumbai and member, Board of Studies of 9 Autonomous Colleges affiliated to Mumbai University. She is member, Academic Council of 2 autonomous college affiliated to University of Mumbai. She is a Special Invitee nominated by the Hon. Vice-chancellor, Faculty of Commerce and Management Dr. Homi Bhabha State University, Mumbai. Seven students are awarded Ph.D under her guidance from University of Mumbai. Eleven students are awarded M. Phil under her guidance from University of Mumbai.

She has organised 31 conferences, workshops and webinars. She has chaired technical session in 36 International/ National conferences/ seminars. She has been Resource person for 145 International/ National / State level events. She has acted as Keynote speaker/ Guest for 35 Conferences/ Seminars. She has 41 publications in International/ National Journals to her credit. She has presented papers in 70 International/ National Conferences/ Seminars. She has co-authored 11 books.

Dr. Shraddha Mayuresh Bhome is a Principal of J.K. College of Science and Commerce, Ghansoli. She has a teaching experience of 18 years (UG), 9 years (PG) and 8 years of Research experience. She has studied M.com, M.Phil in Commerce, Ph.d in commerce. So far 15 Students have successfully completed Ph.D under her guidance and 5 are pursuing Ph.D from Shri JJT University, Rajasthan. She has published 3 patents so far with Govt of India. Written 60+ text books for various universities for undergraduate students, 3 for post graduate students and 3 reference books also to her credit. She has authored, co-authored and presented more than 100 research papers in national, international conferences of which 50+ are awarded as best research papers. She is recipient of 10+ state, national, international awards for the contribution in education and research. Invited as a resource person in more than 60+ colleges across India to deliver various sessions in Research Methodology. She is on the editorial board of 5 journals. Currently, she is Member Board of Studies at 5 colleges. Recently successfully completed GST Practioner Course from IIM Skills with 89%. At present she is Principal of J. K. College of Science and Commerce, Ghansoli.

Financial Aspects in Agriculture

Compiled by

Prashant S. Bodake
Dean & Director of Instruction, DBSKKV
Co-ordinator & Principal Investigator
NAHEP-IDP, DBSKKV-Dapoli, Maharashtra

Srinivasan Iyengar
Director of Jamnalal Bajaj Institute of Management Studies
Mumbai, Maharashtra

Yogesh R. Parulekar
Co-Chair, HORTIC-CITTL
NAHEP-IDP, DBSKKV-Dapoli, Maharashtra

Authors

Kinnarry Thakkar
Professor & Head
Department of Commerce
University of Mumbai, Maharashtra

Shraddha Mayuresh Bhome
Principal
J.K. College of Science and Commerce, Ghansoli

भाकृअनुप
ICAR

Indian Council of Agricultural Research (ICAR)

Dr. Balasaheb Sawant Konkan Krishi Vidyapeeth (DBSKKV)

NAHEP

National Agricultural Higher Education Project-Institutional Development Plan (NAHEP-IDP)

nipa

NIPA® Genx Electronic Resources & Solutions P. Ltd., New Delhi-110 034

Citation: National Agriculture Higher Education Project Institutional Development Plan, DBSKKV, Dapoli & ICAR, New Delhi

NIPA® GENX ELECTRONIC RESOURCES & SOLUTIONS P. LTD.

101,103, Vikas Surya Plaza, CU Block
L.S.C.Market, Pitam Pura, New Delhi-110 034
Ph : +91 11 27341616, 27341717, 27341718
E-mail: newindiapublishingagency@gmail.com
Website: www.nipabooks.com

For customer assistance, please contact
Phone: + 91-11-27 34 17 17
Fax: + 91-11-27 34 16 16

ISBN: 978-93-58871-06-7

Composed and Designed by NIPA®.

भारतीय कृषि अनुसंधान परिषद
कृषि अनुसंधान भवन–II, नई दिल्ली–110012
INDIAN COUNCIL OF AGRICULTURAL RESEARCH
KRISHI ANUSANDHAN BHAVAN-II, PUSA, NEW DELHI-110012
Phone: +91-11-25841760
Email: ddg.edu@icar.gov.in, ddgedn@gmail.com, nd.nahep@icar.gov.in

Dr. Rakesh Chandra Agrawal
Deputy Director General (Agril. Edn.)

Message

It is with immense pleasure and a sense of purpose that I provide this forward to the National Agricultural Higher Education Project-Institutional Development Plan (NAHEP-IDP) at Dr. Balasaheb Sawant Konkan Krishi Vidyapeeth (DBSKKV) and the concurrent endeavour to develop a short pilot course in skilling small-scale agri-entrepreneurship. This transformative initiative is under the guidance and leadership of the Hon'ble Vice Chancellor of DBSKKV Dr.S.G.Bhave.

Principal Investigator of NAHEP-IDP, DBSKKV-Dapoli Dr. P.S.Bodake, a distinguished and visionary leader in the field of agriculture, has consistently championed the cause of advancing agricultural education, research, and extension services. His strategic vision has played a pivotal role in shaping the trajectory of agricultural development in our nation, with a keen focus on empowering farming communities and fostering innovation.

The NAHEP-IDP of DBSKKV stands as a testament to ICAR & World Bank commitment to elevating the standards of agricultural education. By providing institutional support and promoting excellence across various facets of academia, research, and extension activities, the project aims to position DBSKKV as a center of excellence in agricultural learning and research.

In tandem with the NAHEP-IDP, the initiation of a short pilot course in skilling smallscale agri-entrepreneurship reflects DBSKKV foresight in addressing contemporary challenges in agriculture. Recognizing the pivotal role played by small-scale agri-entrepreneurs in driving rural economies, this course is designed to equip them with the necessary skills and knowledge to navigate the dynamic landscape of agribusiness.

The NAHEP-IDP, DBSKKV envisions this pilot course as a stepping stone toward fostering a culture of entrepreneurship in agriculture. By blending theoretical insights with practical applications, the course seeks to empower agri-entrepreneurs, enabling them to adopt sustainable farming practices, embrace innovation, and contribute significantly to the economic prosperity of their communities.

I extend my heartfelt appreciation to Dr. P.S. Bodake and their his NAHEP-IDP Team for their steadfast leadership and unwavering commitment to advancing agricultural education and entrepreneurship. Under his guidance, these initiatives are poised to make a lasting impact on the agricultural education landscape and the entrepreneurial spirit within the students and farming community.

(R.C. Agrawal)
Deputy Director General (Agril. Edn.)

DR. BALASAHEB SAWANT KONKAN KRISHI VIDYAPEETH
Dapoli, Dist. Ratnagiri - 415712
Tel. No.: (0)02358-282064, **Fax No**.: 02358-282074, **Mob.**: 9422556565
E-mail : vcdbskkv-mh@gov.in, ipmbhave@gmail.com, **website** : www.dbskkv.org

Dr. Sanjay G. Bhave
IDP Leader
Vice Chancellor, DBSKKV

Message

I am delighted to present the Institutional Development Plan as part of the National Agricultural Higher Education Project (NAHEP) undertaken by our esteemed institution. This initiative marks a significant milestone in our commitment to advancing agricultural education, research, and innovation, with the ultimate goal of transforming the agricultural landscape of our nation.

This pilot course serves as a testament to our university's dedication to fostering innovation and fostering collaboration between academia and industry. We believe that by offering this course, we are not only contributing to the skill development of individuals but also playing a pivotal role in shaping the future of agriculture.

We, at DBSKKV are happy to felicitate the proficient attainment of the Principal Investigator, IDP Coordinator, and Dean and Director of Instruction of DBSKKV-Dr.Prashant S. Bodake and team NAHEP-IDP involved in the same vertical. Your magnificent work makes all of us highly gratified and prosperous. Well Done!

I extend my gratitude to the faculty members, industry experts, and stakeholders who have contributed their expertise and insights to the development of this course. Their collective efforts have ensured that the curriculum aligns with the ever-changing demands of the agricultural sector.

I am excited to witness the impact that this short pilot course on skilling small-scale agricultural entrepreneurship will have on our participants and, consequently, on the agricultural landscape. With this initiative, we are sowing the seeds of empowerment and innovation, cultivating a new generation of agripreneurs who will not only navigate challenges but also unearth opportunities for growth.

(Sanjay G. Bhave)
IDP Leader
Vice Chancellor, DBSKKV

NATIONAL AGRICULTURAL HIGHER EDUCATION PROJECT
AGRICULTURAL EDUCATION DIVISION
INDIAN COUNCIL OF AGRICULTURAL RESEARCH
(Department of Agricultural Research and Education
Ministry of Agriculture and Farmers Welfare, Government of India)
Room No. 514, Krishi Anusandhan Bhawan-II, Pusa, New Delhi-110 012

Date : 14.12.2023

Dr. N.K. Jain
National Coordinator (IDP)

Message

It is a matter of great pleasure to learn that Dr. Balasaheb Sawant Konkan Krishi Vidyapeeth, Dapoli has developed short pilot course on "Skilling Small-scale Agri-Entrepreneurship" for the undergraduate students. The efforts made by DBSKKV, Dapoli and the Knowledge Experts is a significant stride towards fostering interdisciplinary learning and preparing students for the evolving landscape of agriculture. Agriculture, the backbone of our nation, is undergoing a transformative phase that demands a comprehensive understanding of not only traditional farming practices but also entrepreneurial skills. In this context, DBSKKV, Dapoli is driven by the vision to equip undergraduate students with the knowledge and skills necessary to navigate the complexities of modern agriculture and contribute to the burgeoning field of agri-entrepreneurship. I commend the leadership of Dr. S.G. Bhave, Hon'ble Vice Chancellor, DBSKVV, Dapoli for his forward-thinking approach in embracing this initiative and commitment to provide a robust academic environment aligns seamlessly with our objective of preparing students for the dynamic and competitive landscape of agriculture and entrepreneurship.

I also congratulate Dr. P.S. Bodake, Principal Investigator of NAHEP-IDP, DBSKKV, Dapoli and his entire team, whose collaborative efforts have been instrumental in shaping the curriculum of this course. Together, we envision a program that not only imparts knowledge but also nurtures a spirit of inquiry and enterprise among the students. I extend my warmest congratulations to the University on the launch of the innovative short pilot course on "Skilling Small-scale Agri-Entrepreneurship". May this endeavour be a beacon of progress, fostering excellence in agricultural education and empowering the next generation of agri-entrepreneurs. University's commitment to bridging theory and practical is commendable, and I am confident that this initiative will contribute significantly to the growth of our agricultural landscape.

Place: New Delhi

(N. K. Jain)

NATIONAL AGRICULTURE HIGHER EDUCATION PROJECT INSTITUTIONAL DEVELOPMENT PLAN
DR. BALASAHEB SAWANT KONKAN KRISHI VIDYAPEETH,
DAPOLI, DIST. RATNAGIRI
Mob. No.: 9420413255 / 8275201130
Email.: (P)prameghash@gmail.com (O) hodagrodapoli@gmail.com (O) piidpdbskkv@gmail.com (O)nodalcelldbskkv@gmail.com

Dr. P. S. Bodake
Dean & Director of Instruction
Coordinator, Principal Investigator, NAHEP-IDP

Message

It is with immense pride and anticipation that I present the Institutional Development Plan as part of our active involvement in the National Agricultural Higher Education Project (NAHEP). This plan embodies our institution's unwavering commitment to redefining agricultural education and research, and it signifies a transformative journey towards excellence and innovation.

Through this plan, we aim to enhance our institutional capacities, bolster research initiatives, and cultivate strategic partnerships with industries and communities. By incorporating modern pedagogical techniques, fostering interdisciplinary collaboration, and harnessing the potential of emerging technologies, we strive to create a learning ecosystem that empowers our students to become leaders and change-makers in the agricultural sector and it will not only elevate our institution's stature but also propel the agricultural education and research landscape towards a brighter and more sustainable future.

Our joint commitment to education and empowerment has culminated in the creation of this pilot course. Through a harmonious blend of theoretical insights and practical experiences, participants will delve into market trends, value chain analysis, sustainable practices, financial management, and the integration of technology into agriculture. This holistic approach is designed to prepare individuals to navigate the intricacies of agripreneurship with confidence.

This pilot course is an embodiment of our dedication to fostering collaboration between academia and industry. By bridging the gap between theory and practice, we aim to nurture a new generation of agripreneurs who are poised to steer the agricultural sector towards innovation and growth.

We sincerely mention primitive gratitude towards the former Vice-chancellor of DBSKKV Dr. Sanjay D. Sawant and Vice-chancellor of DBSKKV Dr. Sanjay G. Bhave for such prudent leadership and providence. We feel proud to highlight the sheer hard work and great dedication of the Co-Principal Investigator of NAHEP-IDP and IDP Staff.

We are grateful to the Director of Jamnalal Bajaj Institute of Management-Mumbai Dr. Shrinivasan Iyengar for such excellent association and cordial congruence in the course of developing a short pilot course in skilling small-scale Agri-Entrepreneurship. We also express our sincere gratitude to the faculty members of JBIMS & University of Mumbai, industry experts, and stakeholders whose expertise and contributions have shaped this course. Their collective wisdom ensures that the curriculum remains attuned to the evolving demands of the agricultural landscape.

Thank you.

(P. S. Bodake)
Dean & Director of Instruction
Coordinator, Principal Investigator, NAHEP-IDP

Preface

Agricultural finance is a critical component of sustaining India’s agricultural sector, encompassing the provision of funds and financial services. This subject delves into the meaning, scope, and significance of agricultural finance, highlighting its role in meeting the diverse credit needs of Indian farmers. The sector relies on a mix of institutional and non-institutional sources, including commercial banks, microfinance institutions, rural credit banks, the Lead Bank Scheme, and Regional Rural Banks (RRBs). Recent developments in agricultural credit emphasize technological integration, policy initiatives, and the evolution of sustainable finance practices, reflecting a dynamic approach to addressing the financial requirements of farmers and promoting the overall growth and resilience of Indian agriculture.

This subject provides an introduction to higher financing institutions, focusing on key entities such as the Reserve Bank Of India (RBI), National Bank of Agriculture and Rural Development (NABARD),Asian Development Bank (ADB),International Monetary Fund(IMF), and the World Bank. It explores the roles and functions of these institutions in shaping the financial landscapes, particularly in the context of India. Shifting focus to agricultural cooperation, the subject delves into the meaning and offers a concise historical overview of agricultural cooperation. It further examines the objectives, principles, and the significant role cooperation plays in fostering sustainable agricultural practices, rural development, and the overall well-being of farming communities.

This subject provides insights into the preparation and analysis of financial statements, covering essential components such as balance sheets and income statements. It explores the role of bank norms and the loan process, emphasizing their significance in financial decision-making. Additionally, the subject touches on the concept of the time value of money, elucidating its impact on financial statement analysis and strategic financial planning.

Shifting focus to crop insurance, the subject outlines the concept's meaning and importance, highlighting key features and acknowledging inherent limitations. A specific emphasis is placed on the Pradhan Mantri Fasal Bima Yojana, examining its structure and contributions to mitigating risks in the agricultural sector.

Finally, the subject presents a case study of Amul, a renowned dairy cooperative. It offers a detailed examination of Amul's financial strategies, management practices, and its transformative role in the Indian dairy industry, providing valuable insights into successful cooperative models and sustainable business practices.

Authors

Contents

1

Introduction to Agricultural Credit

The word "Credit" is derived from a Latin word "Credo", meaning "I Believe". The Latin verb "credere" means "to repose confidence in". Note that borrowing is a function of ability to command capital or services currently with a promise to repay it in future i.e. obtaining certain amounts of money as loan to be rapid as specified in the Agreement between the concerned parties. Credit means the ability to command other peoples' capital in return for a promise to repay at some specified time in future. It is therefore the combination of the "ability to borrow" and "willingness to borrow".

In agricultural business credit plays a prominent part particularly in developing countries. Once agricultural credit was supposed to be like a Hangsman's rope but at that time agricultural business was on traditional level without scientific management.

Now, the conditions have very much changed with the development of technology and knowledge spread about the use of new inputs and their management credit gives control over the capital needs of the farmer.

Generally, in farming the fixed capital investment is huge which has come from father to son through inheritance but the problem arises for the running costs. For high income an optimum combination of land, labour, capital and organization is important but if any of these is limited it will tell on the farming efficiency particularly when the capital is short.

The 5Cs of Credit

The 5Cs of Credit that must be considered in lending are:

(i) Character

(ii) Capacity

(iii) Capital

(iv) Condition

(v) Credit worthiness

(i) Character The term "Character" implies credit characters related to the qualities of individuals which make him conscious of the debt obligations. These characters include the borrowers' moral characters like honesty, integrity, sense of responsibility and trustworthiness. It is one of the basic cornerstones in assessing the risk-bearing ability of the borrower. Borrowers with highly rated credit character can withstand unforeseen events and save themselves insolvency. A borrower noted for timely repayment shows a reflection of an ideal credit character. Character is also correlated with the returns and repayment capacities of the borrowers.

(ii) Capacity Shows the capacity of the borrower to pay his debts as at when due. Since payments often depend in part on income, the capacities of borrowers to pay will depend on the income rather than on savings.

(iii) Capital Capital refers to the equity or net worth of a farm business. It assures that funds are available to repay loans if character , capacity prove to be inadequate. Capital also represents one of the cornerstones for measuring the risk bearing ability of the borrower.

(iv) Condition Also signifies the financial conditions of the borrower which has direct relevance with the risk bearing ability as well.

(v) Credit worthiness.

Definition of Agricultural Credit

Agricultural credit is considered as one of the most basic inputs for conducting all agricultural development programmes. In India, there is an immense need for proper agricultural credit as Indian farmers are very poor. From the very beginning, the prime source of agricultural credit in India was moneylenders.

Agricultural Credit could be defined as **"transfer of funds from the have's to have-not's, either directly or indirectly."**

The direct transfer comes in the case of private lending agencies like money lender from the village or otherwise, relatives, friends, big land holders. The indirect transfer of funds is from the institutional agencies such as nationalized commercial banks, cooperative agricultural credit societies, land development banks, regional rural banks, National Agricultural Bank for Agricultural and Rural Development (NABARD).

In the institutional agencies like banks the surplus income is saved for getting the interest which the bank lends to those who borrow to invest for profitable business. In the modern system of farming the role of institutional agricultural credit has become highly important which has been facilitated by the socialization of commercial banks who were very shy in investing in agricultural sector since 1970's.

Now, the question that arises is how should the farmer use agricultural credit so that the business of farming becomes awarding and meets the twin objectives of farm management in profit maximization and family welfare.

In the modern business world the farmer must take decisions for the use of agricultural credit that is, the decision

(a) When to borrow,

(b) How much to borrow,

(c) What to borrow for,

(d) Length of loan and repayment schedule,

(e) What to use for security.

According to agricultural finance experts there are five C's and three R.s.

The three R's are—return from the proposed investment, repaying capacity, and risk bearing ability. The five C's are—character, capacity, capital collateral, conditions and common sense. If the farm operator adheres to the C's and R's and the concepts behind these he would be successful in using the agricultural credit and keep himself out of debts and with prosperity around him.

Length of Agricultural Credit Loan

The time period of return ability are

1. **Short Term Loan or Crop Loan:** It varies from six months to a year. The money is borrowed to support crop production.
2. **Medium Term Loan:**The duration is from two years to five years. This is used for adding assets like buying work stock, poultry farming, buying implements, dairy or cattle fanning. The loan is paid out in annual installment.
3. **Long Term Loan:** This is invested in bringing permanent improvement on the farm like improvement of land, fencing of the farm, land reclamation—these all increase the income generating capacity on the farm. The loan is paid out up to twenty five years period.

This loan is termed as 'land improvement or development loan'. This requires collaterization i.e., Generally, the length of loan should correspond to the length of time it takes to recover the borrowed capital investment with some margin of safety.

The repayment schedule should fit in the flow of income from the invested loan. Long term loan should be amortized (payment made cither in annual and biannual installments. Moreover, the principle gets tapered as the payment is paid and the interest is charged on the current principal).

The surplus accumulated should be kept aside and invested where its opportunity cost is high. Risk must be considered along with profit. If there is unfavorable price or weather condition repayment of loan may be distributed.

Under such conditions the loan should be adjusted to the fluctuating income and weather conditions. One way is to adjust payments by preponement or another way is to adjust interest and principal payment to fluctuating farm income or the length of the loan be increased.

Long-term Agricultural Credit System

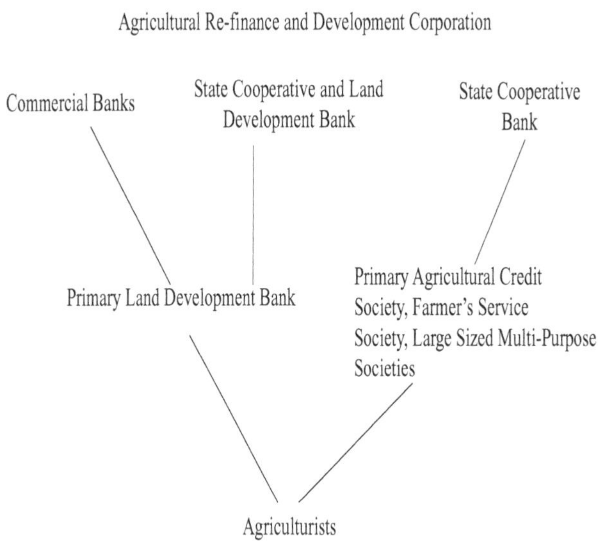

Security for Getting Loan

In case of short term or crop loan no security is required except honesty, managerial ability and promptness in repayment. But in case of medium and long term loans security in terms of the hypothecation of the asset purchased with the bank is a requirement and in case of land development bank loan hypothecation of land is a must.

The Basis of Agricultural Credit

It is not that easy to borrow from the banks as the bank- men are interested in the safe utilization of loans in safeguard of their own interest (loss of fund) and the interest of the client (who should reap good return from the investment of bank loan—supervised credit).

The bank men look for

(a) Farmer's own capital position;

(b) his management ability (they ask for farmer's three years business performance) and his (c) honesty.

The 3 R's & five C's are kept in mind.

Basis for Agricultural Credit

In the case of collateral agricultural credit the farmer need to increase their security base quickly. In case of non-collateral agricultural credit the requirement that the bank manager demand are: managerial ability of the borrower, his honesty, promptness in repayment of loans and his net worth.

Sound Financial Planning for Agricultural Credit

The sound financial planning is the sine quo none of the loan. The farmer should prepare the plan and budget it for calculating the profitability from the investment. In order to have the profitability timely and adequacy of loan is a must because delay in operation costs in terms of low output.

In case, there is an outstanding loan (short term) it should be included in budgeting. The opportunity cost principle and risk considerations are the fundamental guides in the use of borrowed and own capital.

Commercial banks give loans to agriculturists either directly or through Primary Agricultural Credit Society. Or the Regional Rural Banks which give the loan directly to agriculturists or to them through the primary agricultural credit society—the loan which it gets from the commercial banks. The Reserve bank of India either gives loan to regional rural banks which goes to agriculturists through the primary agricultural credit societies or directly to agriculturist from their source.

OR

The Reserve Bank of India passes credit to State Cooperative Bank which passes to Central Cooperative Banks and CCB passes it to Primary Agricultural Credit Societies which ultimately gives loan to agriculturists.

The Agricultural Refinance and Development Corporation gives loan to commercial banks which gives loan to agriculturists. The Agricultural Refinance and Development Corporation passes the loan to State Cooperative and Land Development Bank which then passes the loan to Primary Land Development Bank which ultimately passes it to Agriculturists.

The Agricultural Refinance and Development Corporation gives loan to State Cooperative Banks which passes it on to Primary Agricultural Credit Societies or Farmer Service Societies or Large Sized Multi-purpose Society and finally it reaches the Agriculturists.

Agricultural credit is disbursed through multi-agency network consisting of Commercial Banks, Regional Rural Banks and Cooperatives. The total quantum of agricultural credit from various institutional sources has increased from Rs. 7,005 crores in 1985-86 to Rs. 13,000 crores in 1992-93. The target for agricultural credit disbursement for 1994-95 was Rs. 16,700 crores.

Sources of Rural Credit

Simply understanding what Rural Credit means is not enough. Commerce students also need to learn the various sources from which such monetary assistance is available to rural families. Listed below are the five major sources for Rural Credit in India.

1. Land Development Banks

These banks provide a considerable sum of money as a credit to farmers by using their land as collateral. This low-interest loan has a repayment tenure ranging between 15 and 20 years. Farmers are free to avail this loan to bear the cost of land development work, including the creation of wells or other irrigation related facilities.

Still, land development credits are underutilized since most farmers remain unaware of this source of funding.

2. Co-operative Credit Societies

One of the most economical sources of funding for farmers, co-operative credit facilitates credit to small- and medium-scale farmers. These short-term credits are extended by Primary Agricultural Credit societies or PACs. Nonetheless, these societies have not been able to minimize the influence of moneylenders on the Rural Credit market.

3. Regional Rural Banks

Set up by the government, regional rural banks or RRBs extend monetary assistance to marginal farmers, landless laborers and artisans.

4. Commercial Banks

Originally, commercial banks were reluctant to provide credit for agriculture due to the risks involved with such a move. However, today, these banks extend monetary help both directly and indirectly, to farmers. Direct investment in agriculture refers to short and medium term loans to simplify farming activities. Indirect investment, on the other hand, refers to the advances to farmers made through intermediary agencies or institutions.

5. Government

Also known as Taccavi loans, these are short-term credits extended by the Indian government to assist struggling farmers, especially in the aftermath of natural calamities, such as floods and droughts.

The shares of these agencies are as, cooperatives short term loans 57.5%, commercial banks 37.5%. Regional rural banks 5%. In medium-term and long-term (investments) credit cooperatives account 30% commercial banks 65%, and regional rural banks 5%. The statistics of these lending agencies are: The primary agricultural credit societies 88,000 and primary units for long-term agricultural credit are 2,258. They reached credit to remote corners of the country.

Though there has been reasonably good achievement in cooperative and commercial bank lending which has gone up from 3 per cent and less than 3%, respectively from 1952-54 (as reported in All India Rural Credit Survey) to the level as mentioned, there are many problems faced by these institutions.

There are problems of over dues in cooperatives and RRB's. The government has initiated certain measures to revitalize the cooperatives on the recommendations of Agricultural Credit Review Committee. These measures include amendments of State Cooperative Laws, holding elections of cooperative bodies and revitalizing primary agricultural credit societies through business development planning.

The government is also contemplating the introduction of comprehensive scheme of revitalizing the cooperative agricultural credit structure. NABARD has also initiated the process of drawing up Development Action Plans (DAP's) for cooperative agricultural credit institutions structure in order to make it viable and self-sustaining. In October, 1994, the RBI deregulated the interest rate structure for cooperative lending (subject to a minimum of 12%) and for raising deposits.

Types of Rural Credits

Short Term Credit – These loans have a limited repayment tenure that can range up to one year at the most. Therefore, such credits can act as a brief business or private capital requirement for farmers and others in a rural setting.

Medium Term Loan – Any loan that has a tenure ranging from two years to less than 10 years is classified as a medium-term loan. The credit amount available varies from one firm or individual to the next, depending on the credit rating and a host of other factors.

Long Term Loan– These are considerable sums that farmers can avail for a tenure ranging between 5 years and 20 years. In agriculture, such a line of credit is useful in creating permanent assets. For example, with the help of such a loan, farmers can purchase tractors and other farming properties.

Problems Regarding Agricultural Credit in India

Insufficiency: In spite of the expansion of rural credit structure, the volume of rural credit in the country is still insufficient as compared to its growing requirement arising out of the increase in prices of agricultural inputs.

Inadequate amount of sanction: The amount of loan sanctioned to the farmers by the agencies is also very much inadequate for meeting their different aspects of agricultural operations. Considering the amount of loan sanctioned as inadequate and insignificant, the farmers often divert such loan for unproductive purposes and thereby dilute the very purpose of such loan.

Lesser attention of poor farmers: Rural credit agencies and its schemes have failed to meet the needs of the small and marginal farmers. Thus, lesser attention has been given on the credit needs of the needy farmers whereas the comparatively well-to-do farmers are getting more attention from the credit agencies for their better creditworthiness.

Inadequate institutional coverage: In India, the institutional credit arrangement continues to be inadequate as compared to its growing needs. The development of co-operative credit institutions like Primary agricultural credit societies, land development banks, commercial banks and regional rural banks, have failed to cover the entire rural farmers of the country.

Red tapism: Institutional agricultural-credit is subjected to red-tapism. Credit institutions are still adopting cumbersome rules and formalities for advancing loan to farmers which ultimately force the farmers to depend more on costly non-institutional sources of credit.

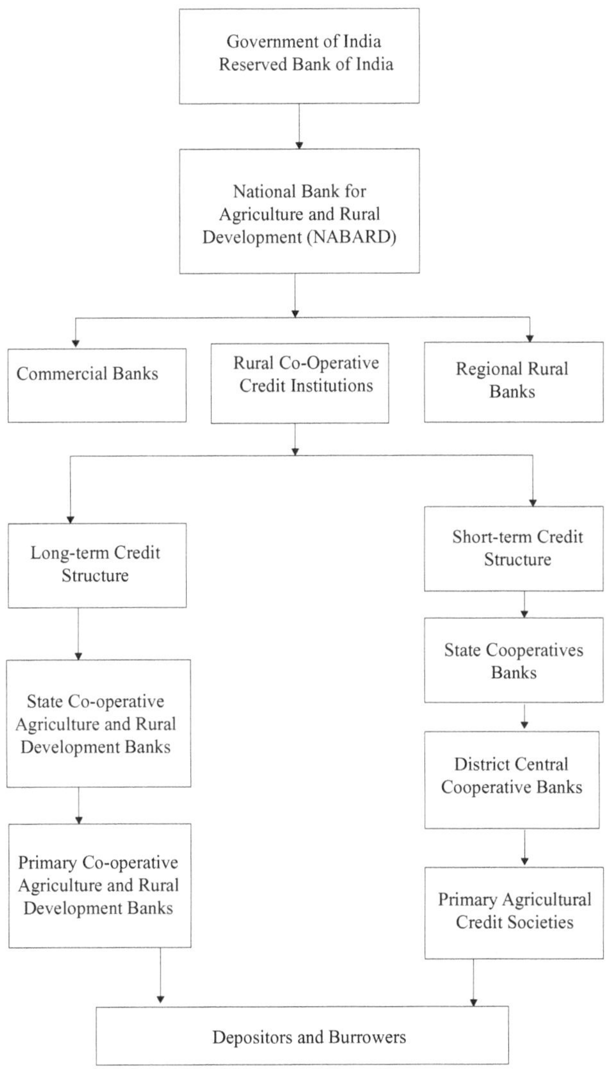

Government Steps for Providing Agriculture Credit to Farmers

• Kisan Credit Card (1998-99)

The Kissan Credit Card (KCC) scheme was launched in 1998 with the aim of providing short-term formal credit to farmers. Owner cultivators, as well as tenant farmers, can avail loans to meet their agricultural needs under this scheme at attractive rates of interest. RBI monitors it for SCBs and NABARD monitors the scheme with respect to Cooperative Banks and RRBs. Now Cooperative sector are under RBI. Budget 2018-19 extended this provision to Animal Husbandry and Fisheries.

• Agriculture Market Infrastructure Fund (AMIF) – NABARD

For development and upgradation of rural agriculture markets. It was announced in 2018 Budget for developing and upgrading agricultural marketing infra in the 22,000 Gramin Agricultural Markets (GrAMs) and 585 APMCs. At present, GrAMs are being developed by MGNREGA Funds. Scheme is demand driven. It will be created with NABARD and will provide the state/ UT governments subsidized loans for their proposal for developing marketing infrastructure in 585 APMCs and 10,000 villages.

• Interest subvention scheme

The Interest Subvention Scheme is being implemented by NABARD and RBI. The interest subvention scheme for farmers aims at providing short term credit to farmers at the subsidized interest rate. The policy came into force with effect from Kharif 2006-07. The scheme is being implemented for the year 2018-19 and 2019-20.

• NABARD Fund of Rupees 700 crore VCF for Rural Agriculture Start ups

The fund has been launched by Nabventures, a subsidiary of NABARD, and has a proposed corpus of Rs 500 crore with an option to retain over-subscription of Rs 200 crore, called as the greenshoe option (overallotment option).

2

Definition of Agricultural Finance Nature-Scope, Meaning-Micro & Macro Finance

Meaning

Agricultural finance generally means studying, examining and analyzing the financial aspects pertaining to farm business, which is the core sector of India. The financial aspects include money matters relating to production of agricultural products and their disposal.

Definition of Agricultural Finance

Murray (1953) defined agricultural. finance as "an economic study of borrowing funds by farmers, the organization and operation of farm lending agencies and of society's interest in credit for agriculture ."

Tandon and Dhondyal (1962) defined agricultural. finance "as a branch of agricultural economics, which deals with and financial resources related to individual farm units."

Nature and Scope

Agricultural finance studied at both micro and macro level. Macrofinance deals with different sources of raising funds for agriculture as a whole in the economy. It is also concerned with the lending procedure, rules, regulations, monitoring and controlling of different agricultural credit institutions. Hence macro-finance is related to financing of agriculture at aggregate level.

Micro-finance refers to financial management of the individual farm business units. And it is concerned with the study as to how the individual farmer considers various sources of credit, quantum of credit to be borrowed from each source and how he allocates the same among the alternative uses with in the farm. It is also concerned with the future use of funds.

Therefore, macro-finance deals with the aspects relating to total credit needs of the agricultural sector, the terms and conditions under which the credit is available and the method of use of total credit for the development of agriculture, while micro-finance refers to the financial management of individual farm business.

Significance of Agricultural Finance

1) Agril finance assumes vital and significant importance in the agro – socio – economic development of the country both at macro and micro level.
2) It is playing a catalytic role in strengthening the farm business and augmenting the productivity of scarce resources. When newly developed potential seeds are combined with purchased inputs like fertilizers & plant protection chemicals in appropriate / requisite proportions will result in higher productivity.
3) Use of new technological inputs purchased through farm finance helps to increase the agricultural productivity.
4) Accretion to in farm assets and farm supporting infrastructure provided by large scale financial investment activities results in increased farm income levels leading to increased standard of living of rural masses.
5) Farm finance can also reduce the regional economic imbalances and is equally good at reducing the inter–farm asset and wealth variations.
6) Farm finance is like a lever with both forward and backward linkages to the economic development at micro and macro level.
7) As Indian agriculture is still traditional and subsistence in nature, agricultural finance is needed to create the supporting infrastructure for adoption of new technology.
8) Massive investment is needed to carry out major and minor irrigation projects, rural electrification, installation of fertilizer and pesticide plants, execution of agricultural promotional programmes and poverty alleviation programmes in the country.

Input finance for purchasing farm materials and livestock funding

For those wondering "What is agriculture finance?" one of the aspects of it is input finance. It is used for purchasing farm materials such as livestock, feed, seeds, and fertiliser. You can also use it to purchase livestock such as pigs, cattle, sheep or goats.

This finance may not cover the costs of large equipment or machinery but it does allow you to purchase other essential items that help with the smooth running of your farm. You can also use this finance to purchase other materials or even small equipment such as compact tractors or farming tools.

Emerging farmer finance

This type of finance is aimed specifically at SMEs and start-up farms. Emerging farmers often battle to gain access to the financial aid that they need, due to the fact they are considered to be higher-risk clients than their more established counterparts.

Emerging farmer finance is usually provided by financial institutions and by commercial banks, as private lenders may view it as a risky investment. If an SME or emerging farmer can be assisted to receiving a contract from a large-scale retailer, this contract can be ceded to a financial institution as security for their loan. You may find that finance for a start-up loan requires some form of collateral.

Asset finance

If you do not have the cash upfront to pay for an important asset, you can look into asset finance to aid the situation. Asset finance requires detailed descriptions of the equipment you need as well what you will be using it for.

Asset finance will allow you to purchase or replace any equipment you may need, such as milking equipment, tractors, harvesting machinery and any other assets that come with a large upfront cost. You could also look into leasing this equipment, if you do not need to use it for an extended period of time.

Establishment loans

An establishment loan is a lon provided to farmers who plant and harvest perennial crops. These loans are available for establishing sugar cane plantations, citrus and deciduous fruit orchards, timber plantations and vineyards for table and wine grapes.

You are not limited with this type of agricultural loan, and can apply for it if you are a commercial farmer, an individual farmer or any legal entity. It can be useful for those who are looking to expand what they will be farming or are interested in other avenues of income regarding agriculture.

Installment finance

This is a form of medium-term loan wherein the goods that you purchase are used as the main security for the loan. This means that the goods belong to the bank until you have repaid the loan and this type of loan is typically used for the purchase of equipment, implements, vehicles or livestock.

Installment finance allows farmers with limited budgets and assets to grow their business. It is the ideal choice for an SME or emerging farmer to choose for purchasing new essential farming equipment. The package is usually available from three to ten years, depending on the expected lifespan of the equipment and any type of farmer may apply for this finance.

Vehicle finance

While this may not seem like the most immediate choice of finance for a farm, the fact of the matter is that farmers need vehicles too. You may need specialist vehicles for certain tasks such as tractors, feed mixers and combine harvesters.

As a farmer you will also need a sturdy vehicle to traverse your farmland, and may not have the extra finances to purchase one. You can choose between hire purchase and lease agreements, and decide on one with better interest rates to suit your needs. This is perfect for farmers who need a new truck or vehicle to transport goods and workers around the farm.

Special mortgage loan

The special mortgage loan is aimed at farmers who have the potential to become successful but who have experienced setbacks such as being denied the right to purchase land. It offers a special interest rate for those who may not be able to pay the higher rates that some banks offer.

The rate is fixed over 24 months and the maximum amount is R500 000. You can pay this loan back over 25 years, making it highly affordable for those who are previously disadvantaged. You will qualify for this loan if you have been denied the right to buy land, are the son or daughter of someone who owns farmland (as a first time buyer) or if you own land in a town or a city.

Conclusion

There are several different options for those who are in need of agricultural finance. You will be able to find one to best suit your needs, and your budget. Be sure to research every aspect of whichever one you choose and read the fine print before signing anything. It may seem a daunting prospect but you will be able to find finance to help your farm grow.

Sources of Agricultural Finance In India

Institutional Sources

- Cooperative Credit societies (Primary Agricultural credit societies, Central Co-operative Bank, state Co-operative Bank)

- Commercial Bank
- Regional Rural Banks (RRB)
- NABARD

Non-Institutional Sources

Money lenders, Traders, Landlords

Co-operative credit societies

This is the cheapest source, these provide short term as well as long term loans.

- Primary Agricultural credit societies at village level
- Central co-operative Bank at District level
- State co-operative Bank at state level
- Commercial Banks

Initially the contribution of commercial bank was very less. After nationalization of banks in 1969, contribution of banks to agricultural credit increased significantly. It provides Direct & Indirect finance. Direct finance for purchasing of pump sets, tractors, other agricultural machinery, construction of wells, tube wells and for other agricultural activities to farmers. Indirect finance is granted to co-operative societies, FCI, state Govt. and other agencies engaged in procurement, storage and distribution of food grains. It also provide credit through service units for warehousing, processing , marketing, transporting etc.

Achievements of Commercial Banks

Share in total rural credit increased from 0.9 % to 74.5%. Helped rural population to free themselves from the cluches of Money lenders. Helped farmers to use the modern methods of cultivation and improve their financial position.

Problems

Huge loss due to branch expansion Cost of service increased with large no. of small borrowings. Problem of bad debt Lack of coordination between the commercial banks, cooperative banks and RRB.

Regional Rural Banks (RRBs)

5 RRBs set up in 1975 for weaker section of rural community. They give direct loans only to small and marginal farmers, agricultural laborers, rural artisans

and small entrepreneurs to develop industry, trade and other productivity in rural areas. They are sponsored by a commercial bank which contributes to 35% of their share capital. Area of operation is limited to a specific region and few districts. Lending rates are low. Refinanced by NABARD.

Achievements of RRBs

Helped needy poor people by providing loan at lowest cost. No. of RRBs increased from 5 to 196. In 2010-11 RRBs provided about 10% of the Institutional credit to agricultural sector.

Problems: Controlled by many agencies central Govt., state Govt. and commercial banks. leads to delay in decision making. Recovery position is poor Substantial loss Lack of proper management. Defective lending policy.

NABARD

National Bank for Agriculture and rural development (NABARD) set up in 1982. it is a refinancing agency. During 2010-11 sanctioned Rs. 35,273 cr. Functions:

- As an apex institution it has to take care of financial requirement of agriculture and rural development.
- Provide short term, medium term and long term credit to cooperatives, RRBs and commercial banks which are main sources of institutional credit.
- It gives long term loan up to 20 years to state govt. to enable them to subscribe to the share capital of cooperative credit societies.
- Direct and supervise the flow of credit to agriculture, a small scale industries, cottage and village industries, handicrafts and other economic activities in rural areas.
- It maintains a Research and Development Fund to promote research in agriculture and rural development.

The Rural Infrastructure Development Fund(RIDF) was established in 1995-96 to finance infrastructure projects. The RIDF gives loans to state Governments for infrastructure projects such as irrigation, rural roads, rural bridges, watershed management, flood control, warehouses, cold storages, fisheries, forest development etc.

NABARD provides 100% refinance assistance to banks at low interest rates for financing self help groups.

It has introduced the Kisan Credit Card scheme in 1998-99 to provide short term credit to farmers. The scheme is implemented by commercial banks, RRBs and cooperatives. Today KCC has become very popular among farmers.

NABARD helps the RRBs and the co-operative banks to implement the Swarnajayanti gram Swarozgar Yojana.

It is responsible for conducting inspection of the cooperatives.

Problems of Agricultural Finanace

Problems of Agricultural Credit in India with Suggested Remedies! Agricultural Credit:

An average Indian farmer, who has to work on an uneconomic holding', using traditional methods of cultivation and being exposed to the risks of a poor agricultural season is almost always in debt. He is a perennial debtor.

Once the farmer falls into debt due to crop failure or low prices of crops or malpractices of moneylenders he can never come out of it. In fact, large part of the liabilities of farmers is 'ancestral debt'. Thus, along with his landed property, he passes on his debt to the next generation.

There are four main causes of rural indebted•ness in India:

(i) low earning power of the bor•rower,

(ii) use of loan for unproductive purposes,

(iii) very high rate of interest charged by the vil•lage moneylender and

(iv) the manipulation of accounts by the lenders.

In a few cases, the bad habits of the farmers (such as gambling, drinking, etc.) are responsible for his burden of 'unproductive' debt. However, in most cases, the cause of the debt may be some expensive social ceremony which the farmer was perhaps forced to "arrange for fear of a social boycott".

Need for Finance

Finance is required by farmers not only for the production and marketing of crops but also to keep a stagnant agricultural economy alive. Most Indian farmers live near the brink of starvation. A bad monsoon, a poor harvest, an accident or illness in the family forces him to approach the moneylender for a loan. In India, there is the preponderance of such 'distress' or unproductive loans. Agricultural finance in India is not just one requirement of the agricultural business but a symptom of the distress prevailing among the majority of the farmers.

Rural credit includes not only credit provided to farmers but also credit extended to artisans, owners of small and medium industries in rural areas, small transport operators and so on. Two main sources of rural credit are private and institutional. The former includes private moneylenders, traders and commission agencies, relatives and- landlords.

The sources of institutional credit are rural co-operatives, commercial banks, particularly the State Bank of India (SBI). And, with the setting up of a specialised institution called the National Bank for Agricultural and Rural Develop•ment (NABARD) the Agricultural Refinance and Development Corporation (ARDC) has ceased to exist. Up to 1982 it was responsible for extending agricultural finance under guidance of the Reserve Bank of India.

It may also be noted that the short- and medium-term credit requirements of the farmers is met by indigenous bankers or village moneylenders, co- operative credit societies and commercial banks. Long-term credit needs are met by land development banks and NABARD.

The principal aim of institutional credit is to replace the widely prevalent money-lending at a very high rate of interest. Available data show that the rural credit institutions have succeeded to a considerable extent in achieving this aim.

Institutional Farm Finance

The need for institutional credit has been felt because of the inherent defects of private agen•cies.

Five main defects of the system of private credit are the following:

1. It is highly exploitative in character because of the inherent profit motive.
2. Since such credit is provided largely for unproductive purposes the rate of interest charged is very high.
3. Such credit is not necessarily directed toward needy persons or desired channels.
4. Such credit is provided for short periods of time and at high rates of interest and cannot, therefore, be utilised for land development or long-term improvement of agriculture.
5. Institutional credit is not linked with other non-farm services such as marketing and processing and warehousing.

By contrast, institutional credit is basically un-exploitive in character. It is largely directed towards raising agricultural productivity so that the income of the farmer increases sufficiently and he becomes self-sufficient. The rate of interest is not only low but varies from case to case. Different rates of interest are charged for different types of loans and different categories of farmers.

Institutional agencies also draw a clear-cut distinction between short-term credit and long- term credit. Moreover, they recognise the organic link between credit and other needs of the farmers and seek to achieve an integration of credit with such needs.

Farmers not only need credit but also guidance in adopting improved methods of cultivation. Thus, it is necessary to provide such guidance and extension services along with credit. They must be taught how to use quality seeds, fertilisers, pesticides, etc. and also how to grow crops.

They must also be provided marketing assistance so that they can obtain the best possible return from their produce. Only institutions like co-operative societies, commercial banks, etc. can provide such guidance, not the usurious moneylenders and greedy commission agents. So it is now necessary to make a brief review of different institutional agencies of rural credit.

Consequences

Rural indebtedness is also likely to have some undesirable social consequences. Due to ever-growing debt there emerges in the rural economy of India a class of landless labourers and tenants. Consequently independent or self-sufficient farmers gradually lose their identity. The landless work•ers have nothing to offer as security in order to obtain loans from moneylenders, except their labour power.

Consequently, they become bonded labourers. This creates discontent among them and adds to rural tensions. In fact, the acquisition of land by the traders and moneylenders and the consequent deprivation of the poor farmers of their meager landed property was the root cause of the Naxalite movement, which assumed serious proportions in West Bengal, Orissa and Andhra Pradesh in the late 1960s and the 1970s. Thus there is no use denying the problem of rural indebtedness. Sooner the problem is removed from its roots the better for India's rural economy.

Suggested Remedies

Since the problem of rural indebtedness has two major dimensions, to solve the problem we have to adopt a two-fold strategy. Since the magnitude of debt is quite high, steps may be taken to cancel old debts. There is a strong case for reduction of ancestral debt and even for their liquidation.

This can be done by State Governments by passing Insolvency Acts. It may be noted that the Government decided in 1990 to write off Rs. 14,000 crores of loans outstanding from farmers Up to a maximum of Rs. 10,000 crores. This was considered necessary because 80% of India's population were farmers and farm workers. Earlier in some States moratorium had been declared on the recovery of debt by moneylenders from farmers, rural artisans and landless workers as per the 20- point Programme.

Secondly, it is to be ensured that the quantum of fresh borrowing is reduced to the minimum, keeping in view the repayment capacity of farmers. It is equally important to ensure that new borrowing is strictly for productive purposes and not for meeting consumption needs. It is, however, difficult for the Government to ensure this in practice.

Only through the spread of education and propaganda among farmers it is possible to check the volume of loans made for unproductive purposes. However, in LDCs like India, arrangement may also be made for providing such loans on a modest scale.

As a subsidiary measure, control of the activities of moneylenders is also necessary. This has been done by some States where sale of land to moneylenders has been prohibited by law.

It is important to note that the abolition of bonded labour and liquidation of rural indebtedness are the two major aspects of the 20-point Programme. The system of bonded labour was abolished by an Act of Parliament in 1976. However, it seemed that the only answer to the present multi-agency credit system is implementing a new multipurpose system with efficient management.

This will have to be so devised as to meet the need for consumption loan of the farmers so that they are not exploited by being paid low wages or low returns on their products. Moreover, the RRBs, if properly managed, can go a long way in solving the problem of rural indebtedness in an effective manner.

Conclusion

Due to extension of institutional credit facilities since 1950-51 the monopoly position of the village moneylender has been challenged. Due to progressive institutionalization of credit, private sources now meet barely 20% of the short- and medium-term credit needs of the farmers.

In other words, institutional sources meet about 80% of such needs. But the Agricultural Credit Review Committee headed by Prof. A. M. Khusro, in its report submitted on August 1989, commented that despite the disappearance of dual financial system, moneylenders are still operating their business in rural India.

One recent study of the Reserve Bank of India admitted that rural households still rely on informal credit markets for 60 to 70% of their credit needs even though interest rates charged are typically over 30%. This is due to apathy of State-owned commercial banks in providing credit to poor peasants. Actually, big landlords are capable of obtaining more loans and advances from various institutions in their own favour at the expense of the poor farmers.

Despite huge increase in overall agricultural credit, there is a serious problem of over-dues which has been inhibiting credit expansion on the one hand and economic viability of the lending institutions, mainly the co-operatives and regional rural banks on the other. The waiver of agricultural loans to the tune of Rs. 10,000 crores in 1990-91 has virtually stopped the credit cycle. If this. practice is continued in the future too rural credit expansion will take a back seat. But, if more and more emphasis is to be given to the agricultural sector, lending institutions will be under more and more pressure.

Above all, small and marginal farmers still remain unworthy borrowers in the banking parlance, though it was hoped at the time of nationalisation that these banks would take care of the credit needs of the farmers. Their dependence on informal markets after 50 years of planning does not augur well. The quantitative expansion of institutional sources hide all these facts. From the qualitative angle, their performance is subject to serious scrutiny.

Role of Cooperative Banks in Agricultural Finance

Though the co-operative credit movement was made a special responsibility of the MI right from the latter's birth in 1935, much was not accomplished in this sphere till about the mid- 1950s. The real turning point in the Bank's role in the movement came only after the Bank's All-India Rural Credit Survey Committee submitted its monumental report in 1954.

The Survey Committee had found that while the co-operative societies and government provided only 3% each of the loans raised by the cultivator, the private credit agencies (the moneylender and the trader) lent more than 70% of what the cultivator borrowed. The moneylender changed very high rates of interest and did not concern himself with the purpose of the loan.

The Survey Committee summed up the position of agricultural credit thus It fell short of the right quantity, was not of the right type, did not serve the right purpose and often failed to go to the right people It also said that 'co- operation had failed but co-operation must succeed'.

For this success, the Survey Committee recommended an 'integrated scheme of rural credit', of which the main features were:

(i) State partnership in co-operative credit institutions through contribution to their share capital;

(ii) Full co-ordination between credit and other economic activi•ties especially marketing and processing; and

(iii) Administration through adequately trained and efficient personnel, responsive to the needs of the rural population.

The RBI was assigned a crucial role in the scheme of integrated credit and in the building up of the co-operative credit organization. The consequent steps taken by the RBI in pursuance of the recommendations of the Survey Committee and later committees like the Committee on Cooperative Credit (1960) transformed the Bank's role from that of a conventional central banker to that of an active agency that takes all necessary measures for enabling the co-operative system to provide a growingly larger share of rural credit.

The adoption of special programmes for increasing agricultural production and the spread of green revolution based largely on intensive use of fertilisers, water, better seeds, and machine power have enhanced the RBI's responsibilities further. The RBI had also started offering greater financial assistance to co-operatives for credit facilities to small farmers and other weaker sections and for minimising disparities in the flow of credit to various regions.

With the setting up of the National Bank for Agriculture and Rural Development (NABARD) in July 1982, the RBI's functions relating to the co- operative movement have been taken over by the NABARD.

Now, the RBI's role is primarily restricted to the provision of finance to the NABARD through its contributions to the two national rural credit funds, already transferred to the NABARD, and additional loans and advances to the latter. Besides, the RBI still offers loans and advances to SCBs.

The NABARD measures are basically a continuation of the RBI measures. They are studied below under two main heads:

(A) Provision of finance and

(B) Building up of the co-operative credit structure.

(A) Provision of Finance

All the NABARD finance is provided to the co-operative sector through the SCBs. The bulk (almost 90%) of it goes to finance agriculture. The finance is of all the three types, viz., short-term, medium-term, and long-term.

(i) **Short-term Agricultural Finance:** This is given primarily for seasonal agricultural operations which are interpreted to include mixed farming activities, i.e., animal husbandry and allied activities jointly undertaken with agricultural operations.

(ii) **Medium-term Agricultural Finance:** The NABARD provides medium-term loans to SCBs for periods of 3 to 5 years. These loans are provided for (a) agricultural purposes (purchase of agricultural machinery, sinking and repair of wells and tube wells, etc.), animal husbandry, poultry farming and for purchase of shares of cooperative sugar factories and other processing societies by agriculturists, and (b) conversion of short-term agricultural loans into medium term loans whenever such conversion becomes necessary on account of wide-spread crop failure as a result of drought, floods or other natural calamities. All medium-term loans are fully guaranteed as to e repayment of the principal and the payment of interest by the state government concerned.

(iii) **Long-term Agricultural Credit:** Long-term credit for agriculture is provided mainly through investment in the debentures of SLDBs. In addition, the National Bank makes long-term loans to state governments for contribution to the share capital of co- operative credit institutions, most of which goes to strengthen co-operative credit for agriculture. The financial accommodation of all kinds indicated above is provided at concessional rates of interest which vary between the Bank Rate and up to 3% below the Bank Rate.

(iv) **Non- agricultural finance**

The NABARD also provides short- term finance for:

(i) The production and marketing activities of selected cottage and small-scale industries (mostly handloom weavers' co-operative societies) and

(ii) The purchase and distribution of fertilisers.

The loans are generally provided through SCBs against guarantees of the state governments. However, all such finance has constituted a small proportion (5 to 7 per cent) of the total Reserve Bank short-term finance to co-operatives: the bulk of it goes to agricultural co-operatives.

During 1994-95, the total amount of financial assistance sanctioned by NABARD was about Rs. 5,300 crore. Of this, about Rs. 4,800 crore were short- term credit and Rs. 500 crore were medium-term credit. The outstanding amount of financial assistance was about Rs. 3,700 crore.

(B) Building up of the Co-operative Credit Structure

From around 1951 the RBI made efforts to (a) strengthen the co-operative credit structure at all the three levels and (b) reorient the operational policies of cooperative banks in more purposive directions. Under the former, the RBI had taken steps to get SCBs established in such states that did not have them and strengthen them where they were weak. The RBI had also tried for the rehabilitation of weak CCBs by prescribing action to recover over dues, strengthen the bad debts reserves and improve the quality of the administrative and supervisory staff.

Similarly, the Bank played an active role in the reorganization of primary societies. The Bank had also made arrangements for the training of personnel of co-operative departments and institutions and undertaken periodical inspection of SCBs, CCBs, and SLDBs to promote healthy and sound growth of co-operative banking in the country. All these functions are now being performed by the NABARD.

Role of RBI in Promoting Commercial Banking, Rural Credit and Industrial Finance!

In addition to performing the traditional functions of a central bank discussed so far, the RBI, after India's independence in 1947,has been playing an active role in two main directions:

(a) In building up and strengthening the country's financial infrastructure, filling up major institutional gaps through the setting up of new financial institutions and reorganizing the existing ones in the context of changing development and other policy needs of the economy and

(b) In devising new measures for influencing the allocation of credit in socially- desired directions. In the discharge of its promotional role, the RBI has several accomplishments to its credit and is continuously engaged in the performance of several non-traditional tasks.

Promotion of Commercial Banking

Under the Banking Regulation Act, 1949, vast powers of supervision and control of commercial banks have been vested in the RBI.

The latter has tried to use these powers:

(a) To strengthen the commercial banking structure in the country through compulsory liquidation of weak banks or their amalgamation into stronger banks and through improvement in the operational standards of banks by regular inspection and general surveillance,

(b) To extend banking facilities throughout the country, especially in small towns and rural areas so as to improve the geographical coverage of banks, and

(c) To extend the functional coverage of banks so as to improve the sectoral distribution of bank credit in favour of the priority sectors such as agriculture, small-scale industries, etc. and make more of it available to small borrowers. The RBI has also arranged for the education and training of different categories of banking personnel.

To inspire greater public confidence in bank deposits and thereby spread banking habit in the country, particularly among the people of small means, insurance of deposits with commercial banks was introduced in January 1962 and a Deposit Insurance Corporation" was set up for the purpose as a subsidiary of the RBI.

In later years the scheme of deposit insurance has been gradually extended to eligible co-operative banks and regional rural banks as well. The amount of deposit eligible tor insurance cover in respect of each depositor in each bank has also been revised upward from Ume to time. Or July 1,990 it was fixed at Rs. 30,000. At the end of June 1995, 70% of total assessable deposits (of Rs. 4, 09,000 crores) of commercial and co-operative banks were insured.

Promotion of Rural (Agricultural) Credit

The provision of adequate amounts of institutional credit for agricultural and other rural activities was recognized as one of the special responsibilities of the RBI even at the time of its birth and appropriate provisions were made to this effect in the Reserve Bank of India Act, 1934, a separate Agricultural Credit Department of the Bank constituted, and the development of co-operative credit movement (which from its inception in 1904 has remained a rural or agricultural movement) made the Bank's special charge.

Much was not done in this sphere till about the mid 50s, when, on the recommendation of the All-India Rural Credit Survey Committee (1954), the (then) Imperial Bank of India and other state-associated banks were nationalised and converted into the State Bank of India and associate banks. This group was made responsible for a vigorous programme of branch expansion in rural areas with a view to provide rural credit as well as mobilise rural savings.

This strategy got a further fillip with the nationalisation of 14 other major commercial banks in July 1969. Other important developments in this field have been the operations of the Agricultural Refinance and Development Corporation (a wholly-owned subsidiary of the RBI, 1963-82); strengthening of the co- operative credit organisation and provision of increasing amount of RBI's refinance to it on concessional terms; setting up of regional rural banks, and channeling of increasing amounts of commercial bank credit to agriculture as a priority sector.

With the setting up of the National Bank for Agriculture and Rural Development in July 1982 to oversee the entire rural credit system and the takeover of the ARDC by it, the direct role and responsibility of the RBI in this sphere have been substantially reduced.

As a result of the measures listed above and increasing emphasis on the provision of institutional finance for agriculture, the picture of such finance has been undergoing rapid changes in favour of agriculture, especially after the nationalisation of 14 major commercial banks in July 1969. We may note for comparison that at the end of March 1995 total gross scheduled commercial bank credit outstanding was Rs. 1, 93,000 crores, about 39% of which had gone to industry.Institutional Finance for Agriculture Promotion of Industrial Finance:

While with some change in the credit policy of commercial banks, the short-term credit needs of large-scale industries could be taken care of relatively easily, the need for special measures was especially acute in two spheres:

(a) The provision of long-term development finance and

(b) Bank credit for small-scale industries.

In both the spheres on the active advice and participation of the RBI special measures have been successfully taken. For providing long-and medium-term finance as well as underwriting of new issues, specialised financial institutions in the form of industrial development banks such as the IDBI, IFCI, ICICI, SIDBI, SFCs and SIDCs have been established in the public sector and the ICICI in the private sector.

The RBI subscribed to the share capital of public sector development banks. It provides them loans from its National Industrial Credit (Long-Term Operations) Fund to which the RBI makes annual contributions from its profits. Started in July 1968 with an initial contribution of only Rs. 10 crore, the Fund had grown to Rs. 5,678 crore on June 30, 1995 and the loans and advances from it stood at Rs. 5,460 crore.

For the small-scale industries, finance is made available by SIDBI, SFCs, and SIDCs and more importantly by commercial banks which are the most important source of credit to them. The recognition of small-scale industries as a 'priority sector' has made all the difference.

At the end of June 1995, credit outstanding-to these industries from public sector banks stood at Rs. 26,800 crore which was about 40% of the total priority sector advances (excluding export credit). Additionally, this finance is provided on concessional terms. One important measure in the promotion of credit to small-scale industries has been the Credit Guarantee Scheme for such industries instituted in 1960 and operated by the RBI on behalf of the Government of India.

Promotion of Export Finance

(i) Various steps have been taken to provide export credit at internationally competitive rates of interest. For example, a scheme was made operative in October 1993 for rediscounting export bills abroad at rates linked to international interest rates. Under another scheme of November 1993, exporters are given pre-shipment credit in major foreign currencies for financing imports.

The RBI provides export credit refinance limits to banks. On March end 1995, they were Rs. 9,400 crore. Export credit refinances limits for post-shipment credit was about Rs. 6,700 crore during 1994-95. Moreover, the rate of interest on export credit has been decontrolled.

The percentage outstanding export credit to net bank credit was 9.3 per cent as on March end 1995. But, the percentage of export credit refinance limits of banks to their outstanding export credit eligible for such refinance was 48 per cent.

(ii) Export-Import Bank

The government has set up in January 1981 an Export-Import Bank, which has taken over the functions of the international financing wing of the IDBI and which acts as the apex institution relating to financing of foreign trade.

Credit to Weaker Sections:

Providing adequate, cheaper, and timely credit to weaker sections is the hardest nut to crack for the policy-makers.

The two measures in this respect taken by the RBI are:

(a) The establishment of the Credit Guarantee Corporation of India in 1971 (merged into the Deposit Insurance Corporation in July 1978) and

(b) The adoption of the Differential Rate of Interest (DRI) Scheme in 1972.

Credit Guarantees

One of the important tasks assigned to the RBI has been to channelise increasing proportion of bank credit in favour of designated priority sectors and small borrowers. Among other things, one major reason for reluctance of banks to provide credit to such priority borrowers has been 'excessive' degree of credit risk involved in lending to them.

Traditional theory would suggest that the free working of market forces will determine appropriate market rate of interest including risk premium for each category of borrowers and that at such rates of interest all borrowers, priority or otherwise, will get as much credit as they would be willing to have. But the market for institutional credit in actual life does not function in this way. The lending rate of interest is officially determined. In such event, high- risk borrowers are simply rationed out and not given credit at higher rates of interest.

An alternative way out is provided by institutional measures that cover the risk of lending agencies. The risk coverage is organized by applying the principle of insurance of pooling individual risks, so that the statistical law of large numbers reduces substantially the credit risk per unit of credit.

This kind of risk coverage has been attempted mainly in the form of credit guarantees. Three separate credit guarantee schemes are now in operation. Their main common feature is to provide guarantees to banks against risk of default of credit extended to designated borrowers eligible, for the guarantee cover, the guarantee cover varying from 66.66% to 100% of the credit under default. The three schemes are explained briefly below.

(i) Credit Guarantee Scheme for Small Scale Industries

To encourage institutional lending to small-scale industries, the Government of India, in consultation with the RBI, introduced a Credit Guarantee Scheme in July 1960 for the guarantee of advances granted by banks and other credit institutions to these industries. The task of administering the Scheme was entrusted to the RBI as the agent of the Central Government.

The scope and provisions of the Scheme have been liberalised from time to time. It extends to all types of credit facilities allowed to small-scale industrial units. The guarantee facilities are available at a small charge to approved credit institutions, including commercial and co-operative banks, regional rural banks, and state financial corporations. Besides, the RBI and the IDBI offer preferential refinance facilities to scheduled commercial banks in respect of short-term lending to small-scale industries covered by the guarantee scheme.

(ii) Credit Guarantee Corporation of India

The Credit Guarantee Scheme for small-scale industries did not solve the problem of meager availability of institutional credit to weaker sections of small transport operators, traders, artisans, self-employed persons, small business enterprises, farmers and agriculturists, etc. Such small borrowers are much less acceptable credit risks to banks than small industries.

Therefore, the RBI, realizing the need for encouraging greater flow of bank credit to small borrowers, established in January 1971 the Credit Guarantee Corporation of India (CGCI) to provide guarantee cover to approved banks in respect of loans and advances to small borrowers. In July 1978 this Corporation was, merged with the Deposit Insurance Corporation which has been renamed as Deposit Insurance and Credit Guarantee Corporation.

(iii) Export Credit and Guarantee Corporation (ECGC)

This Corporation, established by the Government of India in 1964, is under the administrative control of the government, and not of the RBI. As a part of its business is that of issuing credit guarantees. The guarantees are offered to banks and other financial institutions against risk involved in providing export credit, whether in respect of pre-shipment or post-shipment of goods.

The guarantees have been designed to encourage banks to give liberal credit and other facilities for exports. Besides, the ECGC provides insurance covers to exporters against the risk of not receiving payments in respect of export of goods and services.

Differential Rate of Interest (DRI) Scheme

The Scheme has been in operation since 1972. But its progress has been very slow. At the end of June 1995, the advances outstanding under the Scheme from all public sector banks stood at about Rs. 700 crore, constituting 0.33% of their total advances as against the target of 1%.

The advances covered about 23 lakh accounts. About 60% of these advances had gone to members of scheduled castes/scheduled tribes. They showed a very high percentage of over-dues to demand. Such schemes can provide only marginal help to the needy. Much more important are package measures that combine credit facilities with the provision of information, training, inputs, and marketing.

Due to the well-publicised securities-bank scam, 1992 has turned out to be a very trying year for the RBI. The Scam has exposed RBI's poor supervision of banks, and its own public debt office, responsible for recording transactions in

Government of India securities worth thousands of crore rupees per day. Several accusing fingers have been raised against the RBI's demonstrated incompetence, shallow vigilance and the like.Role of RBI in Promoting Commercial Banking, Rural Credit and Industrial Finance!

In addition to performing the traditional functions of a central bank discussed so far, the RBI, after India's independence in 1947,has been playing an active role in two main directions :

(a) In building up and strengthening the country's financial infrastructure, filling up major institutional gaps through the setting up of new financial institutions and reorganizing the existing ones in the context of changing development and other policy needs of the economy and

(b) In devising new measures for influencing the allocation of credit in socially-desired directions. In the discharge of its promotional role, the RBI has several accomplishments to its credit and is continuously engaged in the performance of several non-traditional tasks.

Promotion of Commercial Banking

Under the Banking Regulation Act, 1949, vast powers of supervision and control of commercial banks have been vested in the RBI.

The latter has tried to use these powers:

(a) To strengthen the commercial banking structure in the country through compulsory liquidation of weak banks or their amalgamation into stronger banks and through improvement in the operational standards of banks by regular inspection and general surveillance,

(b) To extend banking facilities throughout the country, especially in small towns and rural areas so as to improve the geographical coverage of banks, and

(c) To extend the functional coverage of banks so as to improve the sectoral distribution of bank credit in favour of the priority sectors such as agriculture, small-scale industries, etc. and make more of it available to small borrowers. The RBI has also arranged for the education and training of different categories of banking personnel.

To inspire greater public confidence in bank deposits and thereby spread banking habit in the country, particularly among the people of small means, insurance of deposits with commercial banks was introduced in January 1962 and a Deposit Insurance Corporation" was set up for the purpose as a subsidiary of the RBI.

In later years the scheme of deposit insurance has been gradually extended to eligible co-operative banks and regional rural banks as well. The amount of deposit eligible tor insurance cover in respect of each depositor in each bank has also been revised upward from Ume to time. Or July 1,990 it was fixed at Rs. 30,000. At the end of June 1995, 70% of total assessable deposits (of Rs. 4, 09,000 crores) of commercial and co-operative banks were insured.

Promotion of Rural (Agricultural) Credit

The provision of adequate amounts of institutional credit for agricultural and other rural activities was recognized as one of the special responsibilities of the RBI even at the time of its birth and appropriate provisions were made to this effect in the Reserve Bank of India Act, 1934, a separate Agricultural

Credit Department of the Bank constituted, and the development of co-operative credit movement (which from its inception in 1904 has remained a rural or agricultural movement) made the Bank's special charge.

Much was not done in this sphere till about the mid 50s, when, on the recommendation of the All-India Rural Credit Survey Committee (1954), the (then) Imperial Bank of India and other state-associated banks were nationalised and converted into the State Bank of India and associate banks. This group was made responsible for a vigorous programme of branch expansion in rural areas with a view to provide rural credit as well as mobilise rural savings.

This strategy got a further fillip with the nationalisation of 14 other major commercial banks in July 1969. Other important developments in this field have been the operations of the Agricultural Refinance and Development Corporation (a wholly-owned subsidiary of the RBI, 1963-82); strengthening of the co-operative credit organisation and provision of increasing amount of RBI's refinance to it on concessional terms; setting up of regional rural banks, and channeling of increasing amounts of commercial bank credit to agriculture as a priority sector.

With the setting up of the National Bank for Agriculture and Rural Development in July 1982 to oversee the entire rural credit system and the takeover of the ARDC by it, the direct role and responsibility of the RBI in this sphere have been substantially reduced.

As a result of the measures listed above and increasing emphasis on the provision of institutional finance for agriculture, the picture of such finance has been undergoing rapid changes in favour of agriculture, especially after the nationalisation of 14 major commercial banks in July 1969. We may note for comparison that at the end of March 1995 total gross scheduled commercial bank credit outstanding was Rs. 1, 93,000 crores, about 39% of which had gone to industry.Institutional Finance for AgriculturePromotion of Industrial Finance:

While with some change in the credit policy of commercial banks, the short-term credit needs of large-scale industries could be taken care of relatively easily, the need for special measures was especially acute in two spheres:

(a) The provision of long-term development finance and

(b) Bank credit for small-scale industries.

In both the spheres on the active advice and participation of the RBI special measures have been successfully taken. For providing long-and medium-term finance as well as underwriting of new issues, specialised financial institutions in the form of industrial development banks such as the IDBI, IFCI, ICICI, SIDBI, SFCs and SIDCs have been established in the public sector and the ICICI in the private sector.

The RBI subscribed to the share capital of public sector development banks. It provides them loans from its National Industrial Credit (Long-Term Operations) Fund to which the RBI makes annual contributions from its profits. Started in July 1968 with an initial contribution of only Rs. 10 crore, the Fund had grown to Rs. 5,678 crore on June 30, 1995 and the loans and advances from it stood at Rs. 5,460 crore.

For the small-scale industries, finance is made available by SIDBI, SFCs, and SIDCs and more importantly by commercial banks which are the most important source of credit to them. The recognition of small-scale industries as a 'priority sector' has made all the difference.

At the end of June 1995, credit outstanding-to these industries from public sector banks stood at Rs. 26,800 crore which was about 40% of the total priority sector advances (excluding export credit). Additionally, this fi•nance is provided on concessional terms. One important measure in the promotion of credit to small-scale industries has been the Credit Guarantee Scheme for such industries instituted in 1960 and operated by the RBI on behalf of the Government of India.

Promotion of Export Finance

(i) Various steps have been taken to provide export credit at internationally competitive rates of interest. For example, a scheme was made operative in October 1993 for rediscounting export bills abroad at rates linked to international interest rates. Under another scheme of November 1993, exporters are given pre-shipment credit in major foreign currencies for financing imports.

The RBI provides export credit refinance limits to banks. On March end 1995, they were Rs. 9,400 crore. Export credit refinances limits for post-shipment credit was about Rs. 6,700 crore during 1994-95. Moreover, the rate of interest on export credit has been decontrolled.

The percentage outstanding export credit to net bank credit was 9.3 per cent as on March end 1995. But, the percentage of export credit refinance limits of banks to their outstanding export credit eligible for such refinance was 48 per cent.

(ii) Export-Import Bank

The government has set up in January 1981 an Export-Import Bank, which has taken over the functions of the international financing wing of the IDBI and which acts as the apex institution relating to financing of foreign trade.

Credit to Weaker Sections

Providing adequate, cheaper, and timely credit to weaker sections is the hardest nut to crack for the policy-makers.

The two measures in this respect taken by the RBI are:

(a) The establishment of the Credit Guarantee Corporation of India in 1971 (merged into the Deposit Insurance Corporation in July 1978) and

(b) The adoption of the Differential Rate of Interest (DRI) Scheme in 1972.

Credit Guarantees

One of the important tasks assigned to the RBI has been to channelise increasing proportion of bank credit in favour of designated priority sectors and small borrowers. Among other things, one major reason for reluctance of banks to provide credit to such priority borrowers has been 'excessive' degree of credit risk involved in lending to them.

Traditional theory would suggest that the free working of market forces will determine appropriate market rate of interest including risk premium for each category of borrowers and that at such rates of interest all borrowers, priority or otherwise, will get as much credit as they would be willing to have. But the market for institutional credit in actual life does not function in this way. The lending rate of interest is officially determined. In such event, high- risk borrowers are simply rationed out and not given credit at higher rates of interest.

An alternative way out is provided by institutional measures that cover the risk of lending agencies. The risk coverage is organized by applying the principle of insurance of pooling individual risks, so that the statistical law of large numbers reduces substantially the credit risk per unit of credit.

This kind of risk coverage has been attempted mainly in the form of credit guarantees. Three separate credit guarantee schemes are now in operation. Their main common feature is to provide guarantees to banks against risk of default of credit extended to designated borrowers eligible, for the guarantee cover, the guarantee cover varying from 66.66% to 100% of the credit under default. The three schemes are explained briefly below.

(i) Credit Guarantee Scheme for Small Scale Industries

To encour•age institutional lending to small-scale industries, the Government of India, in consultation with the RBI, introduced a Credit Guarantee Scheme in July I960 for the guarantee of advances granted by banks and other credit institutions to these industries. The task of administering the Scheme was entrusted to the RBI as the agent of the Central Government.

The scope and provisions of the Scheme have been liberalised from time to time. It extends to all types of credit facilities allowed to small-scale industrial units. The guarantee facilities are available at a small charge to approved credit institutions, including commercial and co-operative banks, regional rural banks, and state financial corporations. Besides, the RBI and the IDBI offer preferential refinance facilities to scheduled commercial banks in respect of short-term lending to small-scale industries covered by the guarantee scheme.

(ii) Credit Guarantee Corporation of India

The Credit Guarantee Scheme for small-scale industries did not solve the problem of meager availability of institutional credit to weaker sections of small transport operators, traders, artisans, self-employed persons, small business enter•prises, farmers and agriculturists, etc. Such small borrowers are much less acceptable credit risks to banks than small industries.

Therefore, the RBI, realizing the need for encouraging greater flow of bank credit to small borrowers, established in January 1971 the Credit Guarantee Corporation of India (CGCI) to provide guarantee cover to approved banks in respect of loans and advances to small borrowers. In July 1978 this Corporation was, merged with the Deposit Insurance Corporation which has been renamed as Deposit Insurance and Credit Guarantee Corporation.

(iii) Export Credit and Guarantee Corporation (ECGC)

This Corporation, established by the Government of India in 1964, is under the administrative control of the government, and not of the RBI. As a part of its business is that of issuing credit guarantees. The guarantees are offered to banks and other financial institutions against risk involved in providing export credit, whether in respect of pre-shipment or post-shipment of goods.

The guarantees have been designed to encourage banks to give liberal credit and other facilities for exports. Besides, the ECGC provides insurance covers to exporters against the risk of not receiving payments in respect of export of goods and services.

Differential Rate of Interest (DRI) Scheme

The Scheme has been in operation since 1972. But its progress has been very slow. At the end of June 1995, the advances outstanding under the Scheme from all public sector banks stood at about Rs. 700 crore, constituting 0.33% of their total advances as against the target of 1%.

The advances covered about 23 lakh accounts. About 60% of these advances had gone to members of scheduled castes/scheduled tribes. They showed a very high percentage of over-dues to demand. Such schemes can provide only marginal help to the needy. Much more important are package measures that combine credit facilities with the provision of information, training, inputs, and marketing.

Due to the well-publicised securities-bank scam, 1992 has turned out to be a very trying year for the RBI. The Scam has exposed RBI's poor supervision of banks, and its own public debt office, responsible for recording transactions in Government of India securities worth thousands of crore rupees per day. Several accusing fingers have been raised against the RBI's demonstrated incompetence, shallow vigilance and the like.

Introduction

In the world of finance and investment, time does have a value. ₹ 100 today is more valuable than ₹ 100 a year later. This is because capital can be employed productively to generate positive returns. Again, individuals normally prefer current consumption to future consumption. Even in the case of inflation, ₹ 100 today represents greater real purchasing power as compared to ₹ 100 one year later. The longer the term of a loan, the greater the amount that must be paid due to interest. Bonds are worthless to an investor, if the maturity is longer. Therefore, this makes sense under the general framework of the time value of money.

Basic Concepts

(a) **Present Value:** A present value is the discounted value of one or more future cash flows.

(b) **Future Value:** A future value is the compounded value of a present value.

(c) **Discounting Factor:** The discounting factor is the present value of a rupee received of a rupee.

(d) **Compounding Factor:** The compounding factor is the future value of a rupee. Discount and compounding factors are functions of two things:

(i) The interest rate used, and

(ii) The time between the present value and the future value. The discount factor decreases as time increases. The discount factor also decreases as interest rate increases.

The time value of money relationship

The basic time values of money relationship are presented in the following equation:

(1) PV = FV × DF

(2) FV = PV × CF

Where

PV = present value

FV = future value

DF = discounting factor = $\frac{1}{(1+R_t)}$

CF = compounding factgor = $(1+R)^t$ R = rate of interest

t = time in years

Future Value of A Single Amount

The future value of an amount invested or borrowed at a given rate of interest can be calculated if the maturity period is given. Suppose a deposit of ₹ 5,000 gets 10% interest compounded annually for a period of 3 years, the future value will be:

PV × CF = 5,000 × (1.10)3 = 5,000 × 1.331 = ₹ 6,655.

Doubling Period

Sometimes, investors should know how long it will take to double his money at a given rate of interest. In this case, a rule of thumb called the rule of 72, can be used. This rule works pretty well for most of the interest rates. The rule of 72 says that it will take 72 years to double your money at 1% interest. You can calculate the doubling period by dividing 72 by interest rate. You can also estimate the interest rate required to double your money in the given number of years by dividing number of years into 72.

For e.g., if the interest rate is 12%, it will take 6 years to double your money (72/12). On the other hand, if you want to double your money in 6 years, the interest rate should be 12% (72/6).

A more accurate method used for doubling your money is using the rule of 69. According to this rule, the doubling period of an investment is $0.35\frac{69}{interestrate}$ thus, the doubling period of interest rate investment of different rates of interest can be determined as follows:

(1) Interest rate 12% = $\frac{69}{12}$ 0.35 + 5.75 = 6.1 years

(2) Interest rate 15% = $\frac{69}{15}$ 0.35 + 5.75 = 6.1 years

Future Value of Annuity

An annuity in the series of payment of a fixed amount for a specified number of periods. When payments are made at the end of each year, it is called ordinary annuity. On the other hand when the payments are made at the beginning of the year, it is called as annuity due. Normally it is assumed that the first annuity payment occurs at the end of the first year.

The future value of an annuity is determined on the basis of the following:

$$\left[FVA = \frac{(1+R)^t}{-1}\right]$$

Where

A = periodic cash payments

= **annual interest rate**

= time in years

the value of $\frac{(1+R)^t}{-1R}$ can be determined by using the time value of money tables.

The future value interest factors (FVIF) for various years are as shown in table:

Year	FVIF @ 8%	FVIF @ 10%	FVIF @ 12%	FVIF @ 14%
1	1.0000	1.0000	1.0000	1.0000
2	2.0800	2.1000	2.2100	2.1400
3	3.2464	3.3100	3.3744	3.4396
4	4.5061	4.6410	4.7793	4.9211
5	5.8666	6.1051	6.3528	6.6101
6	7.3359	7.7156	8.1152	8.5355
7	8.9228	9.4872	10.089	10.730
8	10.636	11.435	12.299	13.232
9	12.486	13.579	14.775	16.085
10	14.486	15.937	17.548	19.337

Present Value

Many times, investors like to know the present value which grows to a given future value. Suppose, you want to save some money from your salary to buy a scooter after 5 years. You should know how much money should be put into bank now in order to get the future value after 5 years. The present value is simply the inverse of compounding between future value and present value is given in the following formula:

where

$$PV = FV \times DF = FVx = \frac{1}{(1+R)^t}$$

Present Value of Uneven Series of Payment

The annuity includes the constant amount in which cash inflows are identical in every period. Many financial decision involve constant cash flows, however, some important decision are concerned with uneven series cash flows. For e.g., investment in shares is expected to pay an increasing series of dividends overtime. The capital budgeting projects also do not normally provide constant cash flows.

In order to deal with uneven payment streams, we have to multiply each payment by the appropriate PVIF and then sum these products to obtain the present value of an uneven series of payments.

Present Value of an Annuity

Many times investors want to know the present value which must be invested today in order to provide an annuity for several future periods. For e.g., a grandfather wants to deposit enough money today to meet the tuition fees of his grandson for the next three years. The interest rate is 8% the present value of this annuity is the sum of the present values of all the future inflows of the annuities. The present value of an annuity can be expressed in the following formula:

$$\left[\frac{(1+R)^t-1}{R(1+R)^t}\right]$$

Where, PVA_t = present value of an annuity with a duration of 't' periods.

A = constant period flow

R = interest rate

The present value interest factor for an annuity (PVIF) can be determined by using the Time Value of Money Tables. The (PVIF) for various year are given below:

Year	FVIF @ 8%	FVIF @ 10%	FVIF @ 12%	FVIF @ 14%
1	0.9259	0.9091	0.8929	0.8772
2	1.7833	1.7355	1.6901	1.6467
3	2.5771	2.4869	2.4018	2.3216
4	3.3121	3.1700	3.0373	2.9140
5	3.9927	3.7908	3.6048	3.4331
6	4.6229	4.3553	4.1114	3.8887
7	5.2064	4.8684	4.5638	4.2883
8	5.7466	5.3346	4.9676	4.6389
9	6.2469	5.7590	5.3282	4.9464
10	6.7101	6.1446	5.6502	5.2161

For all positive interest rates, PVIF for the present value of an annuity is always less than the number of periods the annuity runs, whereas FVIF for the future value of an annuity is equal to or greater than the number of periods.

Illustration 1

The future value of an amount invested or borrowed at a given rate of interest can be calculated if the maturity period is given. Suppose a deposit of ₹ 10,000 gets 10 per cent interest compounded annually for a period of 3 years, the future value will be?

Solution

Given

Present value = 10,000 Compounding Factor = 105 Year = 3 years

To find Future Value FV = PV × CF

$FV = 10,000 \times (1.10)^3$

$= 10,000 \times 1.331$

= ₹13,310

Illustration 2

Satish deposits ₹1,00,000 with a bank which pays 8 per cent interest compounded annually, for a period of 2 years how much he will get at maturity?

Solution

Given

Present Value = 1,00,000 Compounding Factor = 8% To Find FV

$FV = PV \times CF$ $1,00,000 \times (1.08)^2$

$1,00,000 \times 1.1664$

= ₹1,16,640

Illustration 3

Calculate the present value of annuity of ₹ 6,000 received annually for 4 years, when discounting factor is 10%.

Solution

Year	Cash flow	PVIF @ 10%	P.V. (₹)
1	6,000	0.909	54524
2	6,000	0.827	4962
3	6,000	0.751	4506
4	6,000	0.683	4098
		3,170	19,020

Alternate method

Present value factor = 3.170

i.e., $6,000 \times 3.170$ = ₹19,020

Illustration 4

Find out the present value of an annuity of ₹ 30,000 over three years at 10% discount.

Solution

The present value of an annuity of rupee one one received annually for 3 years at 10% is 2.487. Hence present value of an annuity of ₹ 30,000 can be evaluated as

PVA = A × PVIF @ 10%

= 30,000 × 2.487

= ₹74,610

Illustration 5

CSK deposits ₹ 10,000 with a bank at 12% interest compounded quarterly. How much amount he will get after a period of 6 years?

Solution

FV = PV × CF

FV = PV × (1+r / 4)n x 6

= PV × (1 + 0.12 / 4) 4 × 6

= 10,000 × (1.03) 24

= 10,000 × 2.0328

= ₹ 20,328

Illustration 6

Four equal annual payments of ₹ 5,000 are made into a deposit account that pays 8 per cent interest per year. What is the future value of this annuity at the end of 4 years.

Solution

The future value of annuity = ₹5,000 × FVIFA @ 8%

= ₹5,000 × 4.5061

= ₹22,530.50

Illustration 7

City Union Bank advertises that it will pay a lump sum of ₹ 45,740 at the end of 8 years to investors who deposit annually ₹ 4,000 for 8 years. What is the interest rate the bank is paying?

Solution

The interest rate may be calculated in two stages:

a. Find the FVIFA for ₹45,740

$$FVA = A\ (FVIFA)$$

$$45{,}740 = 4{,}000\ (FVIFA)$$

$$FVIFA = \frac{45{,}740}{4{,}000}$$

$$FVIFA = 11.435$$

b. Look at the FVIFA table and read the row corresponding to 8 years until you get the value 11.435. A value equal or close to it is reached and it is at 10%. FVIFA 10%, 8 years = 11.435

Illustration 8

Find the present value of the following cash flow streams. The discount rate is 10 per cent.

Year	Cash flow
1	₹1,000
2	₹4,000
3	₹4,000
4	₹4,000
5	₹3,000

Solution

Year	Payment	PVIF	PV of individual payments
1	1,000	0.9091	909.10
2	4,000	0.8264	3305.60
3	4,000	0.7513	3005.20
4	4,000	0.6830	2732.00
5	3,000	0.6209	1862.70
			11,817.60

Illustration 9

Mr. Mehra has invested ₹ 50,000 on Xerox machine on 1st Jan., 2002. He estimates net cash income from Xerox machine in next 5 years as under:

Year	Estimated inflows
2002	12,000
2003	15,000
2004	18,000
2005	25,000
2006	30,000

At the end of 5th year machine will be sold at scrap value of ₹ 5,000 advice him whether his project is viable, considering interest rate of 10% P.A.

Solution

Calculation of Present Value of Future Cash Flow

Year	Inflows ₹	PVIF @ 10%	PV of inflows
2002	12,000	0.9091	10,909
2003	15,000	0.8264	12,396
2004	18,000	0.7513	13,523
2005	25,000	0.6830	17,075
2006 2006	30,000} 5,000	0.6209	21,732
			75,635

Note: It is assumed that the net cash income is received at the end of the year considering 10% interest rate, the net present value of all future cash flows is ₹75,635 which is greater than the present net cash flow of ₹50,000. Thus, the project is viable.

Illustration 11

XYZ & Co. is considering investing in a project requiring a capital outlay of ₹ 2,00,000. Forecast for annual income after tax is as follows:

Year	1	2	3	4	5
Profit after tax (₹)	1,00,000	1,00,000	80,000	80,000	40,000

Depreciation is 20% on straight line basis

Evaluate the project on the basis of net present value taking 14% discounting factor and advise whether XYZ & Co. should invest in the project or not? The present value of rupee one at 14% discounting rate are 0.8772, 0.7695, 0.6750, 0.5921 and 0.5194.

Solution

Depreciation = 20% of 20,000 = 40,000 Profit after tax is given.

The cash inflow after tax (CFAT) = profit after tax (PAT) + depreciation

Year	PAT	+ Depreciation	CFAT	DF	P.V.
1	1,00,000	40,000	1,40,000	0.8772	1,22,808
2	1.00.000	40,000	1,40,000	0.7695	1,07,730
3	80,000	40,000	1,20,000	0.6750	81,.000
4	80,000	40,000	1,20,000	0.5921	71,052
5	40,000	40,000	80,000	0.5194	41,552
Prersent value of cash inflow					4,24,142
Present value of cash outflow					2,00,000
Net present value					2,24,142

Net present value is positive, hence XYZ & Co. should invest in the project.

Illustration 12

What is the present value of a 4 years annuity of ₹ 5,000 at 12% interest?

Solution

= 8,000 × PVIF @ 12%

= 8,000 × 3.0373 = ₹24.299

Illustration 13

Miss Sonali is considering an investment opportunity which will give her cash inflow of ₹1,000, ₹1,200, ₹1,100 and ₹400 respectively at the end of each of the next 5 years. The initial investment is ₹4,000. If the time, preference rate is 10%, state whether the investment is profitable or not. (present value factor at 10% are 0.9091, 0.8264, 0.7513, 0.6830 and 0.6209)

Solution

Calculation of Present Value

Year	Cash Flow (₹)	Discount Factor	PV (₹)
1	1,000	0.9091	909.10
2	1,500	0.8264	1,239.60
3	1,200	0.7513	901.56
4	1,100	0.6830	751.30
5	400	0.6209	248.39

		PV cash inflow	4,4049.92
		PV cash outflow	4,000.00
		NPV	49.92

As NPV is positive, the investment is profitable

Illustration 14

A 10 payment annuity of ₹ 6,000 will began 7 years. Hence (The first payment occurs at the end of 7th year). What is the present value of this annuity if the discount rate is 14 per cent?

Solution

a. Determine the value of this annuity a year before the first payment begins, i.e., 6 years from now.

$$PVA_t = A\left[\frac{(1+R)^t}{R(1+R)^t}\right]$$

₹6,000 (PVIFA @ 14%, 10) = ₹6000 ×5.2161 = ₹31,296.60

b. Compute the present value of the amount obtained

$$PV = FVX\ \frac{1}{(1+R)^t}$$

= ₹31, 296.60 (PVIF@14%, 6) = 31, 296.60 × 0.4556

= 14, 258.73

Illustration 15

A is due to receive ₹ 10,000 at the end of 5 years. Since A is in need of money immediately, he wants to sell his interest to B. B wants a return of 10% per annum on his investment. How much should he pay A?

Solution

B should pay the present value of ₹ 10,000 at the yield he wants i.e., PV = FV x CF

= 10,000 × (1+ 0.10)5

= 10,000 × (PVIF @ 10%)

= 10,000 × 0.6209

= ₹6,209

Illustration 16

Krishnamurthy has inherited ₹ 1,000 a year for the next 20 years. First payment being made in one year's time. However, he is in need of money immediately and would like to sell his income to a buyer who would pay him the right price. Assume that the current market rate of interest is 9%.

a. What should be the right price he should accept.

b. How much of his income should he sell if he wants only ₹ 2,500 at present.

c. If you were interested in buying the income but, if you had only ₹ 5,000 to invest what would you do?

Solution

a. Present value of annuity

PVA = A × PVIFA @ 9%, 20 years

= ₹1,000 × 9.1285

= ₹9,128.50

Mr. Krishnamurthy should accept ₹9,128.50

b. The present value of rupee one per annum for the next 20 years at 9 per cent per annum is 2,500

₹9.1285. If he wants ₹ 2,500 then he should sell $\frac{2,500}{9.1285} = .87$

Therefore, Mr. Krishnamurthy should sell only ₹ 273.87 of his income, the balance of ₹ 726.13 being kept for himself.

c. You have ₹5,000 to offer to buy his income which is not sufficient to buy it fully. Hence, you can buy part of his income i,e., $\frac{5,000}{9.1285} = 547.74$ and balance of 452.26 will be kept for him

Illustration 17

An investment of ₹40,000 made on 1/04/06 provides inflows as follows:

Date	Alternative I	Alternative II
01/04/06	20,000	10,000
01/04/07	10,000	20,000
01/04/08	10,000	10,000
01/04/09	10,000	10,000

Which alternative you would prefer if the investor's expected return is 10%. Give reasons for your preference.

Solution

Alternative I

Date	Amount	Discount	Factor PV
01/04/06	20,000	0.9091	18,182
01/04/07	10,000	0.8264	8,264
01/04/08	10,000	0.7513	7,513
01/04/09	10,000	0.6830	6830
			40,789

Alternative II

Date	Amount	Discount	Factor PV
01/04/06	10,000	0.9091	9,091
01/04/07	20,000	0.8264	16,528
01/04/08	10,000	0.7513	7,513
01/04/09	10,000	0.6830	6830
			39,962

The net present value of all future cash flows is ₹ 40,789 in case of Alternative I and ₹ 39,962 in case of II. The NPV in case of Alternative I is higher at 10% discounting factor. Hence, Alternative I is preferred.

Illustration 18

The share of Ridhi Ltd., (₹ 10) was quoting at ₹ 102 on 1.04.2002 and the price rose to ₹ 132 on 1.04.2005. Dividends were received at 10% on 30th June each year. Cost of funds was 10% is it worthwhile investment, considering the time value of money (Present value of factor @ 10% were 0.909, 0.826, 0.751)

Solution

Calculation of Present value of Cash inflow:

Year	Inflow₹	Present Value Factor	Present Value
1	1	0.909	0.909
2	1	0.826	0.826
3	1+132 = 133	0.751	99.883
		Present value (-) Present value of Cash flow NPV	101.618
			102.00
			-0.382

Considering the time value of money, the NPV is negative, hence, it is not a wise investment.

Illustration 19

Find out the present value of a debenture from the following:

Face value of Debenture ₹ 1000

Annual Interest Rate 15%

Expected return 12%

Maturity Period 5 years

(Present value of rupee one at 12% are, 0.8929, 0.7972, 0.7118, 0.6355, 0.5674)

Solution

PVd = I (PVAF) + F (DF) = I (PVAF 12%, 5 years) + 5 (DF 12%, 5 years) = 150 (3.6048) + 1,000(0.5674) = 540.72 + 567.40 = 1108.12

Illustration 20

Mr. Vishwanathan is planning to buy a machine which would generate cash flow as follows:

Year	0	1	2	3	4
Cash flow	(25,000)	6,000	8,000	15,000	8,000

If discount rate is 10% is it worth to invest in machine?

Year	1	2	3	4
Discount factor	0.909	0.826	0.751	0.683

Solution

Calculation of NPV

Year	Cash flow	Discount Factor	Present Value
1	6,000	0.909	5,454
2	8,000	0.826	6,608
3	15,000	0.751	11,265
4	8,000	0.683	5,464
Present value of CF			28,791
(-) Present value of outflow			25,000
NPV			3,791

As the NPV is positive, it is worth investing in the machine.

Exercises

1. Indicate the right answer with your reasoning

(a) When compounding is done more than annually, the effective rate of interest is ______________________(i) Greater than the nominal rate of interest

(ii) Lower than nominal rate of interest

(iii) Equal to nominal rate of interest

Ans: (i)

(b) Which provides money with its time value?

(i) Investment (ii) Interest rate

(iii) Market rates (iv) Currency rates

Ans: (ii)

(c) When payments are made at the end of each year, it is known as____ annuity

(i) Annuity due (ii) Ordinary annuity

(iii) Perpetuity (iv) Fixed annuity

Ans: (ii)

2. Ramesh deposited ₹4,000 for 3 years period at 12% interest which is credited at the end of every six months. What will be the total amount credited to Ramesh's account at the end of 3 years?

3. Ganesh plans to send his son for higher studies in America after 5 years. He expects the cost of the study to be ₹4,00,000. How much should he save annually to have a sum of ₹ 4,00,00 at the end of 5 years, if the interest rate is 9%?

4. ICICI Bank promises to give you ₹5,000 after 10 years in exchange of ₹ 2,000 today. What is the interest rate involved in this offer?

5. Arvind wants to invest @ 8% p.a. compound interest, a such amount as will amount to ₹50,000 at the end of three years. How much should he invest?

6. A company has advertised for deposits from the public. If you deposit ₹ 1,000 now, you receive ₹1,464 at the end of 4 years or ₹1,611 at the end of 5 years. What rates of interest is the company paying?

7. Four equal annual payments of ₹ 4,000 are made into a deposit accounts that pays 8 per cent per year. What would be the future value of this annuity at the end of 6 years?

8. You can save ₹ 20,000 a year for 5 years and ₹3,000 a year for 10 years thereafter. What will these savings cumulate to at the end of 15 years if the rate of interest is 10 per cent?

3

Sources of Finance – Financing A New Business

In case of proprietorship business, the individual proprietor generally invests his own savings to start with, and may borrow money on his personal security or the security of his assets from others. Similarly, the capital of a partnership firm consists partly of funds contributed by the partners and partly of borrowed funds. But the company from of organization enables the promoters to raise necessary funds from the public who may contribute capital and become members (share holders) of the company. In course of its business, the company can raise loans directly from banks and financial institutions or by issue of securities (debentures) to the public. Besides, profits earned may also be reinvested instead of being distributed as dividend to the shareholders.

Thus for any business enterprise, there are two sources of finance, viz, funds contributed by owners and funds available from loans and credits. In other words the financial resources of a business may be own funds and borrowed funds.

1. Owner Funds or Ownership Capital

The ownership capital is also known as 'risk capital' because every business runs the risk of loss or low profits, and it is the owner who bears this risk. In the event of low profits they do not have adequate return on their investment. If losses continue the owners may be unable to recover even their original investment. However, in times of prosperity and in the case of a flourishing business the high level of profits earned accrues entirely to the owners of the business. Thus, after paying interest on loans at a fixed rate, the owners may enjoy a much higher rate of return on their investment. Owners contribute risk capital also in the hope that the value of the firm will appreciate as a result of higher earnings and growth in the size of the firm.

The second characteristic of this source of finance is that ownership capital remains permanently invested in the business. It is not refundable like loans or borrowed capital. Hence a large part of it is generally used for a acquiring long—lived fixed assets and to finance a part of the working capital which is permanently required to hold a minimum level of stock of raw materials, a minimum amount of cash, etc.

Another characteristic of ownership capital related to the management of business. It is on the basis of their contribution to equity capital that owners can exercise their right of control over the management of the firm. Managers cannot ignore the owners in the conduct of business affairs. The sole proprietor directly controls his own business. In a partnership firm, the active partner will take part in the management of business. A company is managed by directors who are elected by the members (shareholders).

Arising out of its characteristics, the advantages of ownership capital may be briefly stated as follows:

1. It provides risk capital
2. It is a source of permanent capital
3. It is the basis on which owners acquire their right of control over management
4. It does not require security of assets to be offered to raise ownership capital

There are also certain limitations of ownership capital as a source of finance. These are:

1. The amount of capital, which may be raised as owners fund depends on the number of persons, prepared to take the risks involved. In a partnership confer, a few persons cannot provide ownership capital beyond a certain limit and this limitation is more so in case of proprietary form of organization.
2. A joint stock company can raise large amount by issuing shares to the public. Bus it leads to an increased number of people having ownership interest and right of control over management. This may reduce the original investors' power of control over management. Being a permanent source of capital, ownership funds are not refundable as long as the company is in existence, even when the funds remain idle.
3. A company may find it difficult to raise additional ownership capital unless it has high profit-earning capacity or growth prospects. Issue of additional shares is also subject to so many legal and procedural restrictions.

2. Borrowed Funds and Borrowed Capital

It includes all funds available by way of loans or credit. Business firms raise loans for specified periods at fixed rates of interest. Thus borrowed funds may serve the purpose of long-term, medium-term or short-term finance. The borrowing is generally at against the security of assets from banks and financial institutions. A company to borrow the funds can also issue various types of debentures.

Interest on such borrowed funds is payable at half yearly or yearly but the principal amount is being repaid only at the end of the period of loan. These interest and principal payments have to be met even if the earnings are low or there is loss. Lenders and creditors do not have any right of control over the management of the borrowing firm. But they can sue the firm in a law court if there is default in payment, interest or principal back.

From the business point of view, borrowed capital has several merits.

1. It does not affect the owner's control over management.
2. Interest is treated as an expense, so it can be charged against income and amount of tax payable thereby reduced.
3. The amount of borrowing and its timing can be adjusted according to convenience and needs, and
4. It involves a fixed rate of interest to be paid even when profits are very high, thus owners may enjoy a much higher rate of return on investment then the lenders.

There are certain limitations, too in case of borrowed capacity.

1. Payment of interest and repayment of loans cannot be avoided even if there is a loss. Default in meeting these obligations may create problems for the business and result in decline of its credit worthiness. Continuing default may even lead to insolvency of firm.
2. Secondly, it requires adequate security to be offered against loans. Moreover, high rates of interest may be charged if the firm's ability to repay the loan in uncertain.

Sources of Long Term Finance

Based upon the time, the financial resources may be classified into long term and short term sources of finance. Long term sources of finance are those that are needed over a longer period of time – generally over a year.

A business requires funds to purchase fixed assets like land and building, plant and machinery, furniture etc. These assets may be regarded as the foundation of abusiness. The capital required for these assets is called fixed capital. A part of the working capital is also of a permanent nature. Funds required for this part of the working capital and for fixed capital is called long term finance.

The sources from which a finance manager can raise long-term funds are discussed below:

1. Issue of Shares

The amount of capital decided to be raised from members of the public is divided into units of equal value. These units are known as share and the aggregate values of shares are known as share capital of the company. Those who subscribe to the share capital become members of the company and are called shareholders. They are the owners of the company. Hence shares are also described as ownership securities.

Issue of Equity Shares

The most important source of raising long-term capital for a company is the issue of equity shares. In the case of equity shares there is no promise to shareholders a fixed dividend. But if the company is successful and the level profits are high, equity shareholders enjoy very high returns on their investment. This feature is very attractive to many investors even through they run the risk of having no return if the profits are inadequate or there is loss. They have the right of control over the management of the company and their liability is limited to the value of shares held by them.

From the above it can be said that equity shares have three distinct characteristics:

1. The holders of equity shares are the primary risk bearers. It is the issue of equity shares that mainly provides 'risk capital', unlike borrowed capital. Even compared with preference capital, equity shareholders are to bear ultimate risk.

2. Equity shares enable much higher return sot be earned by shareholders during prosperity because after meeting the preference dividend and interest on borrowed capital at a fixed rate, the entire surplus of profit goes to equity shareholders only.

3. Holders of equity shares have the right of control over the company. Directors are elected on the vote of equity shareholders.

From the company' point of view; there are several merits of issuing equity shares to raise long-term finance.

1. It is a source of permanent capital without any commitment of a fixed return to the shareholders. The return on capital depends ultimately on the profitability of business.
2. It facilities a higher rate of return to be earned with the help borrowed funds. This is possible due to two reasons. Loans carry a relatively lower rate of interest than the average rate of return on total capital. Secondly, there is tax saving as interest paid can be charged to income as a expense before tax calculation.
3. Assets are not required to give as security for raising equity capital. Thus additional funds can be raised as loan against the security of assets.

Although there are several advantages of issuing equity shares to raise long-term capital.

1. The risks of fluctuating returns due to changes in the level of earnings of the company do not attract many people to subscribe to equity capital.
2. The value of shares in the market also fluctuate with changes in business conditions, this is another risk, which many investors want to avoid.

Preference Shares

Preference shares have three distinct characteristics. Preference shareholders have the right to claim dividend at a fixed rate, which is decided according to the terms of issue of shares. Moreover, the preference dividend is to be paid first out of the net profit. The balance, it any, can be distributed among other shareholders that is, equity shareholders. However, payment of dividend is not legally compulsory. Only when dividend is declared, preference shareholders have a prior claim over equity shareholders. Preference shareholders also have the preferential right of claiming repayment of capital in the event of winding up of the company. Preference capital has to be repaid out of assets after meeting the loan obligations and claims of creditors but before any amount is repaid to equity shareholders. Holders of preference shares enjoy certain privileges, which cannot be claimed by the equity shareholders. That is why; they cannot directly take part in matters, which may be discussed at the general meeting of shareholders, or in the election of directors.

Depending upon the terms of conditions of issue, different types of preference shares may be issued by a company to raises funds. Preference shares may be issued as:

1. Cumulative or Non-cumulative
2. Participating or Non-participating

3. Redeemable or Non-redeemable, or as
4. Convertible or non-convertible preference shares.

In the case of cumulative preference shares, the dividend unpaid if any in previous years gets accumulated until that is paid. No cumulative preference shares have any such provision.

Participatory shareholders are entitled to a further share in the surplus profits after a reasonable divided has been paid to equity shareholders. Non-participating preference shares do not enjoy such right. Redeemable preference shares are those, which are repaid after a specified period, where as the irredeemable preference shares are not repaid. However, the company can also redeem these shares after a specified period by giving notice as per the terms of issue. Convertible preference shows are those, which are entitled to be converted into equity shares after a specified period.

Many companies due to the following reasons prefer issue of preference shares as a source of finance.

1. It helps to enlarge the sources of funds.
2. Some financial institutions and individuals prefer to invest in preference shares due to the assurance of a fixed return.
3. Dividend is payable only when there are profits.
4. If does not affect the equity shareholders' control over management

The limitations of preference shares relates to some of its main features:

1. Dividend paid cannot be charged to the company's income as an expense; hence there is no tax saving as in the case of interest on loans.
2. Even through payment of dividend is not legally compulsory, if it is not paid or arrears accumulate there is an adverse effect on the company's credit.
3. Issue of preference share does not attract many investors, as the return is generally limited and not exceed the rates of interest on loan. On the other than, there is a risk of no dividend being paid in the event of falling income.

Methods of Issuing Securities

The firm after deciding the amount to be raised and the type of securities to be issued, must adopt suitable methods to offer the securities to potential investors. There are for common methods followed by companies for the purpose.

- When securities are offered to the general public a document known as Prospectus, or a notice, circular or advertisement is issued inviting the public to subscribe to the securities offered thereby all particulars about the company and the securities offered are made to the public. Brokers are appointed and one or more banks are authorized to collect subscription. This process is termed as Initial Public Offer (IPO).
- Some times the entire issue is subscribed by an organization known as Issue House, which in turn sells the securities to the public at a suitable time.
- The company may negotiate with large investors of financial institutions who agree to take over the securities. This is known as 'Private Placement' of securities.
- When an exiting company decides to raise funds by issue of equity shares, it is required under law to offer the new shares to the existing shareholders. This is described as rights issue of equity shares. But if the existing shareholders decline, the new shares can be offered to the public.

Sources of Short-Term Finance

Short term finance in business usually refers to the additional money a business requires for doing its business for short terms, which is usually a maximum period of one year. These funds are usually used for day to day operations such as payment of wages, inventory ordering, advertisement expenses and so on.

The major sources of short term finance are discussed below:

1. Trade Credit

Trade credit is a common source of short-term finance available to all companies. It refers to the amount payable to the suppliers of raw materials, goods etc. after an agreed period, which is generally less than a year. It is customary for all business firms to allow credit facility to their customers in trade business. Thus, it is an automatic source of finance. With the increase in production and corresponding purchases, the amount due to the creditors also increases. Thereby part of the funds required for increased production is financed by the creditors. The more important advantages of trade credit as a source of short-term finance are the following:

- It is readily available according to the prevailing customs. There are no special efforts to be made to avail of it. Trade credit is a flexible source of finance. It can be easily adjusted to the changing needs for purchases.

- Where there is an open account for any creditor failure to pay the amounts on time due to temporary difficulties does not involve any serious consequence Creditors often adjust the time of payment in view of continued dealings. It is an economical source of finance.

However, the liability on account of trade credit cannot be neglected. Payment has to be made regularly. If the company is required to accept a bill of exchange or to issue a promissory note against the credit, payment must be made on the maturity of the bill or note. It is a legal commitment and must be honored; otherwise legal action will follow to recover the dues.

2. Bank Loans and Advances

Money advanced or granted as loan by commercial banks is known as bank credit. Companies generally secure bank credit to meet their current operating expenses. The most common forms are cash credit and overdraft facilities.

Under the cash credit arrangement the maximum limit of credit is fixed in advance on the security of goods and materials in stock or against the personal security of directors. The total amount drawn is not to exceed the limit fixed. Interest is charged on the amount actually drawn and outstanding. During the period of credit, the company can draw, repay and again draw amounts with in the maximum limit.

In the case of overdraft, the company is allowed to overdraw its current account up to the sanctioned limit. This facility is also allowed either against personal security or the security of assets. Interest is charged on the amount actually overdrawn, not on the sanctioned limit.

The advantage of bank credit as a source of short-term finance is that the amount can be adjusted according to the changing needs of finance. The rate of interest on bank credit is fairly high. But the burden is no excessive because it is used for short periods and is compensated by profitable use of the funds.

Commercial banks also advance money by discounting bills of exchange. A company having sold goods on credit may draw bills of exchange on the customers for their acceptance. A bill is an order in writing requiring the customer to pay the specified amount after a certain period. After acceptance of the bill, the company can drawn the amount as an advance from many commercial banks on payment of a discount. The amount of discount, which is equal to the interest for the period of the bill, and the balance, is available to the company. Bill discounting is thus another source of short-term finance available from the commercial banks.

3. Short Term Loans from Finance Companies

Short-term funds may be available from finance companies on the security of assets. Some finance companies also provide funds according to the value of bills receivable or amount due from the customers of the borrowing company, which they take over.

4. Bridge Loans

A bridge loan is a type of short-term loan, typically taken out for a period of two weeks to three years pending the arrangement of larger or longer-term financing. It is interim financing for an individual or business until permanent or next-stage financing can be obtained. Money from the new financing is generally used to "take out" (i.e. to pay back) the bridge loan, as well as other capitalization needs.

Bridge loans are typically more expensive than conventional financing to compensate for the additional risk of the loan. Bridge loans typically have a higher interest rate, points and other costs that are amortized over a shorter period, as well as various fees and other "sweeteners" like equity participation by the lender. The lender also may require cross-collateralization and a lower loan-to-value ratio. On the other hand, they are typically arranged quickly with little documentation

Retained Earnings

It is an important source of internal or self-financing by a company. So, retained earnings are the cumulative amount of profit or loss left after paying all the expenses and dividend to the shareholders. The cumulative here means it is a continuing account year after year and doesn't close at the end of the accounting year. The retained earnings are also known as earned surplus, retained capital, plowing back of profits or accumulated earnings.

In order for a company to grow, develop and expand it is ideal to use retained earnings for the accumulation of assets that generate income. The generation of income gives more cash flow to the company which gives means for expansion as well as increase its research and development programs. The increase in cash flow of the company helps to improve financial status and makes it favorable in eyes of investor. Retained earnings is favorable for the companies as issuing of new capital is a lengthy process as well as includes several costs, financial obligation and increased risk.

Like an individual, companies too, set aside a part of their profit to meet future requirements. The portion of profits not distributed among the shareholders but retained and used in business is called retained earnings. It is also referred to as ploughing back of profit. This is one of the important sources of internal financing used for fixed as well as working capital. Retained earnings increase the value of shareholders in case of a growing firm.

Merits

- **Permanent Source:** The moment company retains the money in business, nobody can take that from business which helps to keep the financial structure of company fully flexible and increase the creditworthiness of the company. So, it is a permanent source of finance for the company. Due to the retention of earnings the growth and modernization plans of companies don›t suffer due to lack of finance.
- **No Explicit Cost:** Compared to other sources of finance even equity shares or debt, company have to pay some cost as interest or dividend. There is a cost attached to it, company have to bear but in retained earnings we don't have to pay anything to anybody because it is company's own money.
- **Great Degrees:** The retained earnings do no not dilute the ownership of the company resulting in greater operational freedom and flexibility. Company can utilize in any way the business wants. No terms and conditions are applied for realizing the retained earnings.
- **Absorb losses:** The retained earnings help to incur the unexpected losses and company can continue with its business. The company with retained earnings can face unforeseen contingencies, capital market crisis and other downturns.
- **Increase Market Price of Shares:** When your retained earnings keep on increasing, company's market price also increases. When investor invests in company he analyze the balance sheet and see the net worth of the company. Whenever company is dissolved the retained earnings are distributed amongst the shareholder and because of this scenario the market price increases.

Demerits

- **Dissatisfaction:** There is dissatisfaction among shareholders. They don't like retained earnings higher as it decreases their dividend.
- **Uncertain:** The profit earned by the company keeps on changing from year to year. So, the retained earnings is uncertain source of finance.

- **Opportunity Cost:** Everything has an opportunity cost associated with it.It is equal to the income that the shareholders could have otherwise earned by placing these funds in alternative investments.
- **Over Capitalization:** The excess of retained earnings will tempt management to issue bonus share which will lead to over capitalization in business.

Meaning of Debentures

Debentures are one of the frequently used methods by which a business can procure long-term funds for its initial financial needs or for its subsequent requirements of growth and modernisation. Funds acquired by means of debentures represent debt and its holders are the company's creditors.

In common parlance, debenture is merely a written instrument signed by the company under its common seal, acknowledging the debt due by it to its holders. Through this instrument the company promises to pay a specific amount of money as stated therein at a fixed date in future together with periodic payment of interest to compensate the holders for the use of the funds. features of debentures

- Debentures are instruments of debt, which means that debenture holders become creditors of the company
- They are a certificate of debt, with the date of redemption and amount of repayment mentioned on it. This certificate is issued under the company seal and is known as a Debenture Deed
- Debentures have a fixed rate of interest, and such interest amount is payable yearly or half-yearly
- Debenture holders do not get any voting rights. This is because they are not instruments of equity, so debenture holders are not owners of the company, only creditors
- The interest payable to these debenture holders is a charge against the profits of the company. So these payments have to be made even in case of a loss.

Advantages of Debentures

- One of the biggest advantages of debentures is that the company can get its required funds without diluting equity. Since debentures are a form of debt, the equity of the company remains unchanged.

- Interest to be paid on debentures is a charge against profit for the company. But this also means it is a tax-deductible expense and is useful while tax planning
- Debentures encourage long-term planning and funding. And compared to other forms of lending debentures tend to be cheaper.
- Debenture holders bear very little risk since the loan is secured and the interest is payable even in the case of a loss to the company
- At times of inflation, debentures are the preferred instrument to raise funds since they have a fixed rate of interest

References

Brown, C. D., & Williams, E. S. (Year). "Risk Analysis and Financial Planning in Agricultural Enterprises." Journal of Farm Economics, vol. 42, no. 4, pp. 289-307.

Garcia, M. A., & Smith, P. J. (Year). "The Impact of Climate Change on Agricultural Financial Management: A Global Perspective." Environmental Finance, vol. 18, no. 1, pp. 67-82.

Khan, S. A., & Brown, G. R. (Year). "Sustainable Agriculture Financing: Challenges and Opportunities." Journal of Sustainable Agriculture, vol. 32, no. 4, pp. 401-418.

Kim, H. S., & Lee, J. K. (Year). "Impact of Global Commodity Prices on Agricultural Income: A Cross-Country Analysis." Agricultural Economics, vol. 31, no. 2, pp. 167-182.

Miller, D. J., & White, C. R. (Year). "Credit Access and Smallholder Farmer Financial Inclusion in Developing Countries." Journal of Development Economics, vol. 20, no. 1, pp. 56-72.

Patel, M. R., & Jones, L. K. (Year). "Government Policies and Agricultural Finance: A Case Study of Subsidy Programs." Agricultural Finance Review, vol. 28, no. 3, pp. 215-230.

Shrestha, R. K., & Alavalapati, J. R. (Year). "Economic Valuation of Environmental Services Provided by Agroforestry: A Case Study." Agroforestry Systems, vol. 23, no. 3, pp. 175-191.

Smith, J. A., & Johnson, R. B. (Year). "Financial Management Practices in Modern Agriculture: A Comprehensive Review." Journal of Agricultural Economics, vol. 35, no. 2, pp. 145-167.

Turner, H. R., & Davis, W. F. (Year). "Technological Innovations and Financial Performance in Precision Agriculture." International Journal of Agricultural Management, vol. 39, no. 2, pp. 123-145.

Wang, L., & Li, Y. (Year). "Financial Incentives for Sustainable Farming Practices: A Case Study of Organic Agriculture." Journal of Agricultural and Resource Economics, vol. 26, no. 3, pp. 451-468.

FSC
www.fsc.org
MIX
Papier | Fördert gute Waldnutzung
FSC® C083411

Zeitfracht Medien GmbH
Ferdinand-Jühlke-Straße 7
99095 Erfurt, Deutschland
produktsicherheit@kolibri360.de